Thomas Henry Huxley

Evidence as to Man's Place in Nature

Thomas Henry Huxley

Evidence as to Man's Place in Nature

ISBN/EAN: 9783337026196

Printed in Europe, USA, Canada, Australia, Japan

Cover: Foto ©berggeist007 / pixelio.de

More available books at **www.hansebooks.com**

EVIDENCE

AS TO

MAN'S PLACE IN NATURE.

BY

THOMAS H. HUXLEY, F.R.S., F.L.S.,

PROFESSOR OF NATURAL HISTORY IN THE JERMYN STREET SCHOOL OF MINES.

NEW YORK:

D. APPLETON AND COMPANY,

1, 3, AND 5 BOND STREET.

1886.

CONTENTS.

ADVERTISEMENT TO THE READER.

THE greater part of the substance of the following Essays has already been published in the form of Oral Discourses, addressed to widely different audiences, during the past three years.

Upon the subject of the second Essay, I delivered six Lectures to the Working Men in 1860, and two, to the members of the Philosophical Institution of Edinburgh in 1862. The readiness with which my audience followed my arguments, on these occasions, encourages me to hope that I have not committed the error, into which working men of science so readily fall, of obscuring my meaning by unnecessary technicalities: while, the length of the period during which the subject, under its various aspects, has been present to my mind, may suffice to satisfy the Reader that, my conclusions, be they right or be they wrong, have not been formed hastily or enunciated crudely.

T. H. H.

LONDON: *January,* 1863.

I.

ON THE NATURAL HISTORY OF THE MAN-LIKE APES.

ANCIENT traditions, when tested by the severe processes of modern investigation, commonly enough fade away into mere dreams: but it is singular how often the dream turns out to have been a half-waking one, presaging a reality. Ovid foreshadowed the discoveries of the geologist : the Atlantis was an imagination, but Columbus found a western world: and though the quaint forms of Centaurs and Satyrs have an existence only in the realms of art, creatures approaching man

FIG. 1.—Simiæ magnatum deliciæ.—De Bry, 1598.

more nearly than they in essential structure, and yet as

thoroughly brutal as the goat's or horse's half of the mythical compound, are now not only known, but notorious.

I have not met with any notice of one of these MAN-LIKE APES of earlier date than that contained in Pigafetta's " Description of the kingdom of Congo," * drawn up from the notes of a Portuguese sailor, Eduardo Lopez, and published in 1598. The tenth chapter of this work is entitled " De Animalibus quæ in hac provincia reperiuntur," and contains a brief passage to the effect that " in the Songan country, on the banks of the Zaire, there are multitudes of apes, which afford great delight to the nobles by imitating human gestures." As this might apply to almost any kind of apes, I should have thought little of it, had not the brothers De Bry, whose engravings illustrate the work, thought fit, in their eleventh " Argumentum," to figure two of these " Simiæ magnatum deliciæ." So much of the plate as contains these apes is faithfully copied in the woodcut (fig. 1), and it will be observed that they are tail-less, long-armed, and large-eared; and about the size of Chimpanzees. It may be that these apes are as much figments of the imagination of the ingenious brothers as the winged, two-legged, crocodile-headed dragon which adorns the same plate; or, on the other hand, it may be that the artists have constructed their drawings from some essentially faithful description of a Gorilla or a Chimpanzee. And, in either case, though these figures are worth a passing

* REGNUM CONGO: hoc est VERA DESCRIPTIO REGNI AFRICANI QUOD TAM AB INCOLIS QUAM LUSITANIS CONGUS APPELLATUR, per Philippum Pigafettam, olim ex Edoardo Lopez acroamatis lingua Italica excerpta, num Latio sermone donata ab August. Cassiod. Reinio. Iconibus et imaginibus rerum memorabilium quasi vivis, opera et industria Joan. Theodori et Joan. Israelis de Bry, fratrum exornata. Francofurti, MDXCVIII.

notice, the oldest trustworthy and definite accounts of any animal of this kind date from the 17th century, and are due to an Englishman.

The first edition of that most amusing old book, " Purchas his Pilgrimage," was published in 1613, and therein are to be found many references to the statements of one whom Purchas terms " Andrew Battell (my neere neighbour, dwelling at Leigh in Essex) who served under Manuel Silvera Perera, Governor under the King of Spaine, at his city of Saint Paul, and with him went farre into the countrey of Angola ; " and again, " my friend, Andrew Battle, who lived in the kingdom of Congo many yeares," and who, " upon some quarell betwixt the Portugals (among whom he was a sergeant of a band) and him, lived eight or nine moneths in the woods." From this weather-beaten old soldier, Purchas was amazed to hear " of a kinde of Great Apes, if they might so bee termed, of the height of a man, but twice as bigge in feature of their limmes, with strength proportionable, hairie all over, otherwise altogether like men and women in their whole bodily shape.* They lived on such wilde fruits as the trees and woods yielded, and in the night time lodged on the trees."

This extract is, however, less detailed and clear in its statements than a passage in the third chapter of the second part of another work—" Purchas his Pilgrimes," published in 1625, by the same author—which has been often, though hardly ever quite rightly, cited. The chapter is entitled, " The strange adventures of Andrew Battell, of Leigh in Essex, sent by the Portugals prisoner to Angola, who lived there and in the adioning regions neere eight-eene yeeres." And the sixth section of this chapter is

---

* " Except this that their legges had no calves."—[Ed. 1626.] And in a marginal note, " These great apes are called Pongo's."

headed—" Of the Provinces of Bongo, Calongo, Mayombe, Manikesocke, Motimbas : of the Ape Monster
Pongo, their hunting : Idolatries ; and divers other observations."

" This province (Calongo) toward the east bordereth
upon Bongo, and toward the north upon Mayombe,
which is nineteen leagues from Longo along the coast.

" This province of Mayombe is all woods and groves,
so overgrowne that a man may travaile twentie days in
the shadow without any sunne or heat.   Here is no
kind of corne nor graine, so that the people liveth onely
upon plantanes and roots of sundrie sorts, very good ;
and nuts ; nor any kinde of tame cattell nor hens.

" But they have great store of elephant's flesh, which
they greatly esteeme, and many kinds of wild beasts ;
and great store of fish.   Here is a great sandy bay, two
leagues to the northward of Cape Negro,* which is the
port of Mayombe.   Sometimes the Portugals lade log-
wood in this bay.   Here is a great river, called Banna :
in the winter it hath no barre, because the generall
winds cause a great sea.   But when the sunne hath his
south declination, then a boat may goe in ; for then it is
smooth because of the raine.   This river is very great,
and hath many ilands and people dwelling in them.
The woods are so covered with baboones, monkies, apes
and parrots, that it will feare any man to travaile in
them alone.   Here are also two kinds of monsters, which
are common in these woods, and very dangerous.

" The greatest of these two monsters is called Pongo
in their language, and the lesser is called Engeco.   This
Pongo is in all proportion like a man ; but that he is
more like a giant in stature than a man ; for he is very

---

* Purchas' note.—Cape Negro is in 16 degrees south of the line.

tall, and hath a man's face, hollow-eyed, with long haire upon his browes. His face and eares are without haire, and his hands also. His bodie is full of haire, but not very thicke; and it is of a dunnish colour.

"He differeth not from a man but in his legs; for they have no calfe. Hee goeth alwaies upon his legs, and carrieth his hands clasped in the nape of his necke when he goeth upon the ground. They sleepe in the trees, and build shelters for the raine. They feed upon fruit that they find in the woods, and upon nuts, for they eate no kind of flesh. They cannot speake, and have no understanding more than a beast. The people of the countrie, when they travaile in the woods, make fires where they sleepe in the night; and in the morning when they are gone, the Pongoes will come and sit about the fire till it goeth out; for they have no understanding to lay the wood together. They goe many together, and kill many negroes that travaile in the woods. Many times they fall upon the elephants which come to feed where they be, and so beate them with their clubbed fists, and pieces of wood, that they will runne roaring away from them. Those Pongoes are never taken alive because they are so strong, that ten men cannot hold one of them; but yet they take many of their young ones with poisoned arrowes.

"The young Pongo hangeth on his mother's belly with his hands fast clasped about her, so that when the countrie people kill any of the females they take the young one, which hangeth fast upon his mother.

"When they die among themselves, they cover the dead with great heaps of boughs and wood, which is commonly found in the forest." *

* Purchas' marginal note, p. 982 :—" The Pongo is a giant ape. He told me in conference with him, that one of these Pongoes tooke a negro boy of

It does not appear difficult to identify the exact region of which Battell speaks. Longo is doubtless the name of the place usually spelled Loango on our maps. Mayombe still lies some nineteen leagues northward from Loango, along the coast; and Cilongo or Kilonga, Manikesocke, and Motimbas are yet registered by geographers. The Cape Negro of Battell, however, cannot be the modern Cape Negro in 16° S., since Loango itself is in 4° S. latitude. On the other hand, the "great river called Banna" corresponds very well with the "Camma" and "Fernand Vas," of modern geographers, which form a great delta on this part of the African coast.

Now this "Camma" country is situated about a degree and a half south of the Equator, while a few miles to the north of the line lies the Gaboon, and a degree or so north of that, the Money River—both well known to modern naturalists as localities where the largest of man-like Apes has been obtained. Moreover, at the present day, the word Engeco, or N'schego, is applied by the natives of these regions to the smaller of the two great Apes which inhabit them; so that there can be no rational doubt that Andrew Battell spoke of that which he knew of his own knowledge, or, at any rate, by immediate report from the natives of Western Africa. The "Engeco," however, is that "other monster" whose nature Battell "forgot to relate," while the name "Pongo"—applied to the animal whose characters and habits are so fully and carefully described—seems to have died

his which lived a moneth with them. For they hurt not those which they surprise at unawares, except they looke on them; which he avoyded. He said their highth was like a man's, but their bignesse twice as great. I saw the negro boy. What the other monster should be he hath forgotten to relate; and these papers came to my hand since his death, which, otherwise, in my often conferences, I might have learned. Perhaps he meaneth the Pigmy Pongo killers mentioned."

out, at least in its primitive form and signification. Indeed, there is evidence that not only in Battell's time, but up to a very recent date, it was used in a totally different sense from that in which he employs it.

For example, the second chapter of Purchas' work, which I have just quoted, contains "A Description and Historicall Declaration of the Golden Kingdom of Guinea, &c. &c. Translated from the Dutch, and compared also with the Latin," wherein it is stated (p. 986) that—

"The River Gaboon lyeth about fifteen miles northward from Rio de Angra, and eight miles northward from Cape de Lope Gonsalvez (Cape Lopez), and is right under the Equinoctial line, about fifteene miles from St. Thomas, and is a great land, well and easily to be knowne. At the mouth of the river there lieth a sand, three or foure fathoms deepe, whereon it beateth mightily with the streame which runneth out of the river into the sea. This river, in the mouth thereof, is at least foure miles broad; but when you are about the Iland called *Pongo*, it is not above two miles broad. . . . On both sides the river there standeth many trees. . . . . . The Iland called *Pongo*, which hath a monstrous high hill."

The French naval officers, whose letters are appended to the late M. Isidore Geoff. Saint Hilaire's excellent essay on the Gorilla,* note in similar terms the width of the Gaboon, the trees that line its banks down to the water's edge, and the strong current that sets out of it. They describe two islands in its estuary;—one low, called Perroquet; the other high, presenting three conical hills, called Coniquet; and one of them, M. Franquet, expressly states that, formerly, the Chief of Coniquet was called

* Archives du Museum, Tome X.

Meni-Pongo, meaning thereby Lord of *Pongo;* and that the *N'Pongues* (as, in agreement with Dr. Savage, he affirms the natives call themselves) term the estuary of the Gaboon itself *N'Pongo.*

It is so easy, in dealing with savages, to misunderstand their applications of words to things, that one is at first inclined to suspect Battell of having confounded the name of this region, where his "greater monster" still abounds, with the name of the animal itself. But he is so right about other matters (including the name of the "lesser monster,") that one is loth to suspect the old traveller of error; and, on the other hand, we shall find that a voyager of a hundred years' later date speaks of the name "Boggoe," as applied to a great Ape, by the inhabitants of quite another part of Africa—Sierra Leone.

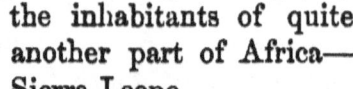
Homo Sylvestris.
Orang Outang.

Fig. 2.—The Orang of Tulpius, 1641.

But I must leave this question to be settled by philologers and travellers; and I should hardly have dwelt so long upon it except for the curious part played by this word '*Pongo*' in the later history of the man-likeApes.

The generation which succeeded Battell saw the first of the man-like Apes which was ever brought to Europe, or, at any rate, whose visit found a historian. In the third book of Tulpius' "Observationes Medicæ," published in 1641, the 56th chapter or section is devoted to what he calls *Satyrus indicus* " called by the Indians Orang-autang, or Man-of-the-Woods, and by the Africans Quoias Morrou." He gives a very good figure, evidently from the life, of the specimen of this animal, " nostra memoria ex Angolâ delatum," presented to Frederick Henry, Prince of Orange. Tulpius says it was as big as a child of three years old, and as stout as one of six years: and that its back was covered with black hair. It is plainly a young Chimpanzee.

In the meanwhile, the existence of other, Asiatic, man-like Apes became known, but at first in a very mythical fashion. Thus Bontius (1658) gives an altogether fabulous and ridiculous account and figure of an animal which he calls " Orang-outang"; and though he says " vidi Ego cujus effigiem hic exhibeo," the said effigies (see fig. 6 for Hoppius' copy of it) is nothing but a very hairy woman of rather comely aspect, and with proportions and feet wholly human. The judicious English anatomist, Tyson, was justified in saying of this description by Bontius, " I confess I do mistrust the whole representation."

It is to the last mentioned writer, and his coadjutor Cowper, that we owe the first account of a man-like ape which has any pretensions to a scientific accuracy and completeness. The treatise entitled, *Orang-outang, sive Homo Sylvestris ;* or the Anatomy of a Pygmie compared with that of a *Monkey*, an *Ape*, and a *Man*," published by the Royal Society in 1699, is, indeed, a work of remarkable merit, and has, in some respects, served as a model to subsequent inquirers. This " Pygmie," Tyson tells us, " was brought from Angola, in Africa ; but was first taken a great deal higher up the country ; " its hair " was of a coal-black colour, and strait," and " when it went as a quadruped on all four, 'twas awkwardly ; not placing the palm of the hand flat to the ground, but it walk'd upon its knuckles, as I observed it to do when weak and had not strength enough to support its body." —"From the top of the head to the heel of the foot, in a straight line, it measured twenty-six inches."

FIGS. 3 & 4.—The 'Pygmie' reduced from Tyson's figures 1 and 2, 1699.

These characters, even without Tyson's good figures (figs. 3 and 4), would have been sufficient to prove his "Pygmie" to be a young Chimpanzee. But the opportunity of examining the skeleton of the very animal Tyson anatomised having most unexpectedly presented itself to me, I am able to bear independent testimony to its being a veritable *Troglodytes niger*,* though still very young. Although fully appreciating the resemblances between his Pygmie and Man, Tyson by no means overlooked the differences between the two, and he concludes his memoir by summing up first, the points in which "the Ourang-outang or Pygmie more resembled a Man than Apes and Monkeys do," under forty-seven distinct heads; and then giving, in thirty-four similar brief paragraphs, the respects in which "the Ourang-outang or Pygmie differ'd from a Man and resembled more the Ape and Monkey kind."

After a careful survey of the literature of the subject extant in his time, our author arrives at the conclusion that his "Pygmie" is identical neither with the Orangs of Tulpius and Bontius, nor with the Quoias Morrou of Dapper (or rather of Tulpius), the Barris of d'Arcos, nor with the Pongo of Battell; but that it is a species of ape probably identical with the Pygmies of the Ancients, and, says Tyson, though it "does so much resemble *a Man* in many of its parts, more than any of the ape kind, or any other *animal* in the world, that I know of: yet by no means do I look upon it as the product of a *mixt* generation—'tis a *Brute-Animal sui generis*, and a particular *species of Ape.*"

The name of "Chimpanzee," by which one of the African Apes is now so well known, appears to have

---

* I am indebted to Dr. Wright, of Cheltenham, whose paleontological labours are so well known, for bringing this interesting relic to my knowledge. Tyson's granddaughter, it appears, married Dr. Allardyce, a physician of repute in Cheltenham, and brought, as part of her dowry, the skeleton of the 'Pygmie.' Dr. Allardyce presented it to the Cheltenham Museum, and, through the good offices of my friend Dr. Wright, the authorities of the Museum have permitted me to borrow, what is, perhaps, its most remarkable ornament.

come into use in the first half of the eighteenth century, but the only important addition made, in that period, to our acquaintance with the man-like apes of Africa is contained in " A New Voyage to Guinea," by William Smith, which bears the date 1744.

In describing the animals of Sierra Leone, p. 51, this writer says :—

" I shall next describe a strange sort of animal, called by the white men in this country Mandrill,* but why it is

FIG. 5.—Facsimile of William Smith's figure of the " Mandrill," 1744.

* " Mandrill" seems to signify a "man-like ape," the word "Drill" or "Dril" having been anciently employed in England to denote an Ape or Baboon. Thus in the fifth edition of " Blount's " Glossographia, or a Dictionary interpreting the hard words of whatsoever language now used in our refined English tongue . . . very useful for all such as desire to understand what they read," published in 1681, I find, " Dril—a stone-cutter's tool wherewith he bores little holes in marble, &c. Also a large overgrown Ape and Baboon, so called." " Drill" is used in the same sense in Charleton's "Onomasticon Zoicon," 1668. The singular etymology of the word given by Buffon seems hardly a probable one.

so called I know not, nor did I ever hear the name before, neither can those who call them so tell, except it be for their near resemblance of a human creature, though nothing at all like an Ape. Their bodies, when full grown, are as big in circumference as a middle-sized man's—their legs much shorter, and their feet larger; their arms and hands in proportion. The head is monstrously big, and the face broad and flat, without any other hair but the eyebrows; the nose very small, the mouth wide, and the lips thin. The face, which is covered by a white skin, is monstrously ugly, being all over wrinkled as with old age; the teeth broad and yellow; the hands have no more hair than the face, but the same white skin, though all the rest of the body is covered with long black hair, like a bear. They never go upon all-fours, like apes; but cry, when vexed or teased, just like children. . . . . .

"When I was at Sherbro, one Mr. Cummerbus, whom I shall have occasion hereafter to mention, made me a present of one of these strange animals, which are called by the natives Boggoe: it was a she-cub, of six months' age, but even then larger than a Baboon. I gave it in charge to one of the slaves, who knew how to feed and nurse it, being a very tender sort of animal; but whenever I went off the deck the sailors began to teaze it—some loved to see its tears and hear it cry; others hated its snotty-nose; one who hurt it, being checked by the negro that took care of it, told the slave he was very fond of his country-woman, and asked him if he should not like her for a wife? To which the slave very readily replied, ' No, this no my wife; this a white woman—this fit wife for you.' This unlucky wit of the negro's, I fancy, hastened its death, for next morning it was found dead under the windlass."

William Smith's ' Mandrill,' or ' Boggoe,' as his description and figure testify, was, without doubt, a Chimpanzee.

Linnæus knew nothing, of his own observation, of the man-like Apes of either Africa or Asia, but a dissertation by his pupil Hoppius in the " Amœnitates Academicæ " (VI. ' Anthropomorpha ') may be regarded as embodying his views respecting these animals.

The dissertation is illustrated by a plate, of which the accompanying woodcut, fig. 6, is a reduced copy.   The figures are entitled (from left to right) 1. *Troglodyta Bontii* ; 2. *Lucifer Aldrovandi* ; 3. *Satyrus Tulpii* ; 4. *Pygmæus Edwardi*.   The first is a bad copy of Bontius' fic-

Fɪɢ. 6.—The Anthropomorpha of Linnæus.

titious ' Ourang-outang,' in whose existence, however, Linnæus appears to have fully believed ; for in the standard edition of the " Systema Naturæ," it is enumerated as a second species of Homo ; " H. nocturnus."   *Lucifer Aldrovandi* is a copy of a figure in Aldrovandus, ' De Quadrupedibus digitatis viviparis,' Lib. 2, p. 249 (1645) entitled " Cercopithecus formæ raræ *Barbilius* vocatus et

originem a china ducebat." Hoppius is of opinion that this may be one of that cat-tailed people, of whom Nicolaus Köping affirms that they eat a boat's crew, "gubernator navis" and all! In the "Systema naturæ" Linnæus calls it in a note, *Homo caudatus*, and seems inclined to regard it as a third species of man. According to Temminck, *Satyrus Tulpii* is a copy of the figure of a Chimpanzee published by Scotin in 1738, which I have not seen. It is the *Satyrus indicus* of the "Systema Naturæ," and is regarded by Linnæus as possibly a distinct species from *Satyrus sylvestris*. The last, named *Pygmœus Edwardi*, is copied from the figure of a young "Man of the Woods," or true Orang-Utan, given in Edwards' 'Gleanings of Natural History,' (1758).

Buffon was more fortunate than his great rival. Not only had he the rare opportunity of examining a young Chimpanzee in the living state, but he became possessed of an adult Asiatic man-like Ape—the first and the last adult specimen of any of these animals brought to Europe for many years. With the valuable assistance of Daubenton, Buffon gave an excellent description of this creature, which, from its singular proportions, he termed the long-armed Ape, or Gibbon. It is the modern *Hylobates lar*.

Thus when, in 1766, Buffon wrote the fourteenth volume of his great work, he was personally familiar with the young of one kind of African man-like Ape, and with the adult of an Asiatic species—while the Orang-Utan and the Mandrill of Smith were known to him by report. Furthermore, the Abbé Prevost had translated a good deal of Purchas' Pilgrims into French, in his 'Histoire générale des Voyages' (1748), and there Buffon found a version of Andrew Battell's account of the Pongo and the Engeco. All these data Buffon attempts to weld together into harmony in his chapter entitled "Les Orang-outangs

ou le Pongo et le Jocko." To this title the following note
is appended :—

"Orang-outang nom de cet animal aux Indes orientales: Pongo nom de
cet animal à Lowando Province de Congo.
"Jocko, Enjocko, nom de cet animal à Congo que nous avons adopté.
En est l'article que nous avons retranché."

Thus it was that Andrew Battell's " Engeco " became
metamorphosed into " Jocko," and, in the latter shape,
was spread all over the world, in consequence of the extensive popularity of Buffon's works. The Abbé Prevost
and Buffon between them however, did a good deal more
disfigurement to Battell's sober account than ' cutting off
an article.' Thus Battell's statement that the Pongos
" cannot speake, and have no understanding more than a
beast," is rendered by Buffon " qu'il ne peut parler
*quoiqu'il ait plus d'entendement que les autres ani-
maux ;* " and again, Purchas' affirmation, " He told me
in conference with him, that one of these Pongos tooke a
negro boy of his which lived a moneth with them," stands
in the French version, " un pongo lui enleva un petit
negre qui passa un *an* entier dans la societé de ces ani-
maux."

After quoting the account of the great Pongo, Buffon
justly remarks, that all the ' Jockos ' and ' Orangs ' hitherto brought to Europe were young ; and he suggests that,
in their adult condition, they might be as big as the Pongo
or ' great Orang ; ' so that, provisionally, he regarded the
Jockos, Orangs, and Pongos as all of one species. And
perhaps this was as much as the state of knowledge at the
time warranted. But how it came about that Buffon
failed to perceive the similarity of Smith's ' Mandrill ' to
his own ' Jocko,' and confounded the former with so to-

tally different a creature as the blue-faced Baboon, is not so easily intelligible.

Twenty years later Buffon changed his opinion,[*] and expressed his belief that the Orangs constituted a genus with two species,—a large one, the Pongo of Battell, and a small one, the Jocko : that the small one (Jocko) is the East Indian Orang ; and that the young animals from Africa, observed by himself and Tulpius, are simply young Pongos.

In the meanwhile, the Dutch naturalist, Vosmaer, gave, in 1778, a very good account and figure of a young Orang, brought alive to Holland, and his countryman, the famous anatomist, Peter Camper, published (1779) an essay on the Orang-Utan of similar value to that of Tyson on the Chimpanzee. He dissected several females and a male, all of which, from the state of their skeleton and their dentition, he justly supposes to have been young. However, judging by the analogy of man, he concludes that they could not have exceeded four feet in height in the adult condition. Furthermore, he is very clear as to the specific distinctness of the true East Indian Orang.

" The Orang," says he, " differs not only from the Pigmy of Tyson and from the Orang of Tulpius by its peculiar colour and its long toes, but also by its whole external form. Its arms, its hands, and its feet are longer, while the thumbs, on the contrary, are much shorter, and the great toes much smaller in proportion." [†] And again, " The true Orang, that is to say, that of Asia, that of Borneo, is consequently not the Pithecus, or tail-less Ape, which the Greeks, and especially Galen, have described. It is neither the Pongo nor the Jocko, nor the Orang of

---

[*] Histoire Naturelle, Suppl. tome 7ème, 1789.

[†] Camper, Œuvres, I., p. 56.

2

Tulpius, nor the Pigmy of Tyson,—*it is an animal of a peculiar species*, as I shall prove in the clearest manner by the organs of voice and the skeleton in the following chapters." (l. c. p. 64).

A few years later, M. Radermacher, who held a high office in the Goverument of the Dutch dominions in India, and was an active member of the Batavian Society of Arts and Sciences, published, in the second part of the Transactions of that Society,* a Description of the Island of Borneo, which was written between the years 1779 and 1781, and, among much other interesting matter, contains some notes upon the Orang. The small sort of Orang-Utan, viz. that of Vosmaer and of Edwards, he says, is found only in Borneo, and chiefly about Banjermassing, Mampauwa, and Landak. Of these he had seen some fifty during his residence in the Indies; but none exceeded 2½ feet in length. The larger sort, often regarded as chimæra, continues Radermacher, would, perhaps, long have remained so, had it not been for the exertions of the Resident at Rembang, M. Palm, who, on returning from Landak towards Pontiana, shot one, and forwarded it to Batavia in spirit, for transmission to Europe.

Palm's letter describing the capture runs thus:— " Herewith I send your Excellency, contrary to all expectation (since long ago I offered more than a hundred ducats to the natives for an Orang-Utan of four or five feet high) an Orang which I heard of this morning about eight o'clock. For a long time we did our best to take the frightful beast alive in the dense forest about half way to Landak. We forgot even to eat, so anxious were we not to let him escape ; but it was necessary to take care he did not revenge himself, as he kept continually break-

* Verhandelingen van het Bataviaasch Genootschap. Tweede Deel. Derde Druk. 1826.

ing off heavy pieces of wood and green branches, and dashing them at us. This game lasted till four o'clock in the afternoon, when we determined to shoot him; in which I succeeded very well, and indeed better than I ever shot from a boat before; for the bullet went just into the side of his chest, so that he was not much damaged. We got him into the prow still living, and bound him fast, and next morning he died of his wounds. All Pontiana came on board to see him when we arrived." Palm gives his height from the head to the heel as 49 inches.

A very intelligent German officer, Baron Von Wurmb, who at this time held a post in the Dutch East India service, and was Secretary of the Batavian Society, studied this animal, and his careful description of it, entitled " Beschrijving van der Groote Borneosche Orang-outang of de Oost-Indische Pongo," is contained in the same volume of the Batavian Society's Transactions. After Von Wurmb had drawn up his description he states, in a letter dated Batavia, Feb. 18, 1781,* that the specimen was sent to Europe in brandy to be placed in the collection of the Prince of Orange; " unfortunately," he continues, " we hear that the ship has been wrecked." Von Wurmb died in the course of the year 1781, the letter in which this passage occurs being the last he wrote; but in his posthumous papers, published in the fourth part of the Transactions of the Batavian Society, there is a brief description, with measurements, of a female Pongo four feet high.

Did either of these original specimens, on which Von Wurmb's descriptions are based, ever reach Europe? It is commonly supposed that they did; but I doubt the fact. For, appended to the memoir " De l'Ourang-outang," in the collected edition of Camper's works, Tome I., pp.

---

* " Briefe des Herrn v. Wurmb und des H. Baron von Wollzogen. Gotha, 1794 "

64–66, is a note by Camper himself, referring to Von Wurmb's papers, and continuing thus :—"Heretofore, this kind of ape had never been known in Europe. Radermacher has had the kindness to send me the skull of one

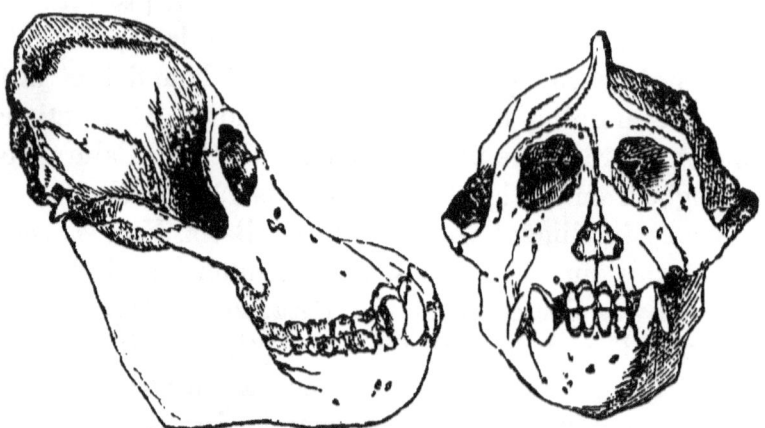

FIG. 7.—The Pongo Skull, sent by Radermacher to Camper, after Camper's original sketches, as reproduced by Lucæ.

of these animals, which measured fifty-three inches, or four feet five inches, in height. I have sent some sketches of it to M. Soemmering at Mayence, which are better calculated, however, to give an idea of the form than of the real size of the parts."

These sketches have been reproduced by Fischer and by Lucæ, and bear date 1783, Soemmering having received them in 1784. Had either of Von Wurmb's specimens reached Holland, they would hardly have been unknown at this time to Camper, who, however, goes on to say :—"It appears that since this, some more of these monsters have been captured, for an entire skeleton, very badly set up, which had been sent to the Museum of the Prince of Orange, and which I saw only on the 27th of June, 1784, was more than four feet high. I examined this skeleton again on the 19th December, 1785, after it had been excellently put to rights by the ingenious Ony-mus."

It appears evident, then, that this skeleton, which is doubtless that which has always gone by the name of Wurmb's Pongo, is not that of the animal described by him, though unquestionably similar in all essential points.

Camper proceeds to note some of the most important features of this skeleton ; promises to describe it in detail by-and-bye ; and is evidently in doubt as to the relation of this great ' Pongo' to his " petit Orang."

The promised further investigations were never carried out ; and so it happened that the Pongo of Von Wurmb took its place by the side of the Chimpanzee, Gibbon, and Orang as a fourth and colossal species of man-like Ape. And indeed nothing could look much less like the Chim-panzees or the Orangs, then known, than the Pongo ; for all the specimens of Chimpanzee and Orang which had been observed were small of stature, singularly human in aspect, gentle and docile ; while Wurmb's Pongo was a monster almost twice their size, of vast strength and fierceness, and very brutal in expression ; its great pro-jecting muzzle, armed with strong teeth, being further disfigured by the outgrowth of the cheeks into fleshy lobes.

Eventually, in accordance with the usual marauding habits of the Revolutionary armies, the ' Pongo' skeleton was carried away from Holland into France, and notices of it, expressly intended to demonstrate its entire distinct-ness from the Orang and its affinity with the baboons, were given, in 1798, by Geoffroy St. Hilaire and Cuvier.

Even in Cuvier's " Tableau Elementaire," and in the first edition of his great work, the " Regne Animal," the ' Pongo' is classed as a species of Baboon. However, so

early as 1818, it appears that Cuvier saw reason to alter
this opinion, and to adopt the view suggested several years
before by Blumenbach,* and after him by Tilesius, that
the Bornean Pongo is simply an adult Orang. In 1824,
Rudolphi demonstrated, by the condition of the dentition,
more fully and completely than had been done by his pred-
ecessors, that the Orangs described up to that time were
all young animals, and that the skull and teeth of the
adult would probably be such as those seen in the Pongo
of Wurmb. In the second edition of the 'Regne Animal'
(1829), Cuvier infers, from the 'proportions of all the
parts' and 'the arrangements of the foramina and sutures
of the head,' that the Pongo is the adult of the Orang-
Utan, 'at least of a very closely allied species,' and this
conclusion was eventually placed beyond all doubt by
Professor Owen's Memoir published in the 'Zoological
Transactions' for 1835, and by Temminck in his 'Monog-
raphies de Mammalogie.' Temminck's memoir is remarkable for the completeness of the evidence which it affords
as to the modification which the form of the Orang undergoes according to age and sex. Tiedemann first published
an account of the brain of the young Orang, while Sandifort, Müller and Schlegel, described the muscles and the
viscera of the adult, and gave the earliest detailed and
trustworthy history of the habits of the great Indian Ape
in a state of nature ; and as important additions have been
made by later observers, we are at this moment better acquainted with the adult of the Orang-Utan, than with that
of any of the other greater man-like Apes.

It is certainly the Pongo of Wurmb ;† and it is as

* See Blumenbach, "Abbildungen Naturhistorichen Gegenstände," No. 12,
1810; and Tilesius, "Naturhistoriche Früchte der ersten Kaiserlich-Rus-
sischen Erdumsegelung," p. 115, 1813.

† Speaking broadly and without prejudice to the question, whether there
be more than one species of Orang.

certainly not the Pongo of Battell, seeing that the Orang-Utan is entirely confined to the great Asiatic islands of Borneo and Sumatra.

And while the progress of discovery thus cleared up the history of the Orang, it also became established that the only other man-like Apes in the eastern world were the various species of Gibbon—Apes of smaller stature, and therefore attracting less attention than the Orangs, though they are spread over a much wider range of country, and are hence more accessible to observation.

Although the geographical area inhabited by the 'Pongo' and 'Engeco' of Battell is so much nearer to Europe than that in which the Orang and Gibbon are found, our acquaintance with the African Apes has been of slower growth; indeed, it is only within the last few years that the truthful story of the old English adventurer has been rendered fully intelligible. It was not until 1835 that the skeleton of the adult Chimpanzee became known, by the publication of Professor Owen's above-mentioned very excellent memoir " On the osteology of the Chimpanzee and Orang," in the Zoological Transactions—a memoir which, by the accuracy of its descriptions, the carefulness of its comparisons, and the excellence of its figures, made an epoch in the history of our knowledge of the bony framework, not only of the Chimpanzee, but of all the anthropoid Apes.

By the investigations herein detailed, it became evident that the old Chimpanzee acquired a size and aspect as different from those of the young known to Tyson, to Buffon, and to Traill, as those of the old Orang from the young Orang; and the subsequent very important researches of Messrs. Savage and Wyman, the American missionary and anatomist, have not only confirmed this conclusion, but have added many new details.*

One of the most interesting among the many valuable discoveries made by Dr. Thomas Savage is the fact, that the natives in the Gaboon country at the present day, apply to the Chimpanzee a name—" Enché-eko "—which is obviously identical with the " Engeko " of Battell ; a discovery which has been confirmed by all later inquirers. Battell's " lesser monster " being thus proved to be a veritable existence, of course a strong   resumption arose that his " greater monster," the ' Pongo,' would sooner or later be discovered.   And, indeed, a modern traveller, Bowdich, had, in 1819, found strong evidence, among the natives, of the existence of a second great Ape, called the ' Ingena,' " five feet high, and four across the shoulders," the builder of a rude house, on the outside of which it slept.

In 1847, Dr. Savage had the good fortune to make another and most important addition to our knowledge of the man-like Apes ; for, being unexpectedly detained at the Gaboon river, he saw in the house of the Rev. Mr. Wilson, a missionary resident there, " a skull represented by the natives to be a monkey-like animal, remarkable for its size, ferocity, and habits." From the contour of the skull, and the information derived from several intelligent natives, " I was induced," says Dr. Savage, (using the term Orang in its old general sense) " to believe that it belonged to a new species of Orang.   I expressed this opinion to Mr. Wilson, with a desire for further investiga-

* See " Observations on the external characters and habits of the Troglodytes niger, by Thomas N. Savage, M. D., and on its organization, by Jeffries Wyman, M. D.," Boston Journal of Natural History, Vol. IV. 1843–4 ; and " External characters, habits, and osteology of Troglodytes Gorilla," by the same authors, ibid. Vol. V. 1847.

tion ; and, if possible, to decide the point by the inspection of a specimen alive or dead." The result of the combined exertions of Messrs. Savage and Wilson was not only the obtaining of a very full account of the habits of this new creature, but a still more important service to science, the enabling the excellent American anatomist already mentioned, Professor Wyman, to describe, from ample materials, the distinctive osteological characters of the new form. This animal was called by the natives of the Gaboon " Engé-ena," a name obviously identical with the " Ingena" of Bowdich ; and Dr. Savage arrived at the conviction that this last discovered of all the great Apes was the long-sought ' Pongo' of Battell.

The justice of this conclusion, indeed, is beyond doubt —for not only does the ' Engé-ena' agree with Battell's " greater monster" in its hollow eyes, its great stature, and its dun or iron-grey colour, but the only other man-like Ape which inhabits these latitudes—the Chimpanzee —is at once identified, by its smaller size, as the " lesser monster," and is excluded from any possibility of being the ' Pongo,' by the fact that it is black and not dun, to say nothing of the important circumstance already mentioned that it still retains the name of ' Engeko' or ' Enché-eko,' by which Battell knew it.

In seeking for a specific name for the ' Enge-ena,' however, Dr. Savage wisely avoided the much misused ' Pongo'; but finding in the ancient Periplus of Hanno the word " Gorilla" applied to certain hairy savage people, discovered by the Carthaginian voyager in an island on the African coast, he attached the specific name " *Gorilla*" to his new ape, whence arises its present well-known appellation. But Dr. Savage, more cautious than some of his successors, by no means identifies his ape with Hanno's ' wild men.' He merely says that the latter

2*

were " probably one of the species of the Orang ; " and I quite agree with M. Brullé, that there is no ground for identifying the modern ' Gorilla' with that of the Carthaginian admiral.

Since the memoir of Savage and Wyman was published, the skeleton of the Gorilla has been investigated by Professor Owen and by the late Professor Duvernoy, of the Jardin des Plantes, the latter having further supplied a valuable account of the muscular system and of many of the other soft parts ; while African missionaries and travellers have confirmed and expanded the account originally given of the habits of this great man-like Ape, which has had the singular fortune of being the first to be made known to the general world and the last to be scientifically investigated.

Two centuries and a half have passed away since Battell told his stories about the ' greater' and the ' lesser monsters' to Purchas, and it has taken nearly that time to arrive at the clear result that there are four distinct kinds of Anthropoids—in Eastern Asia, the Gibbons and the Orangs ; in Western Africa, the Chimpanzees and the Gorilla.

The man-like Apes, the history of whose discovery has just been detailed, have certain characters of structure and of distribution in common. Thus they all have the same number of teeth as man—possessing four incisors, two canines, four false molars, and six true molars in each jaw, or 32 teeth in all, in the adult condition ; while the milk dentition consists of 20 teeth—or four incisors, two canines, and four molars in each jaw. They are what are called catarrhine Apes—that is, their nostrils have a narrow partition and look downwards ; and, furthermore, their arms are always longer than their legs, the difference being sometimes greater and sometimes less ; so that if the four were arranged in the order of the length of their arms in proportion to that of their legs, we should have this series—Orang ($1\frac{4}{5}$—1), Gibbon ($1\frac{1}{4}$—1), Gorilla ($1\frac{1}{5}$—1), Chimpanzee ($1\frac{1}{16}$—1). In all, the fore limbs are terminated by hands, provided with longer or shorter thumbs ; while the great toe of the foot, always smaller than in Man, is far more moveable than in him and can be opposed, like a thumb, to the rest of the foot. None of these apes have tails, and none of them possess the cheek-pouches common among monkeys. Finally, they are all inhabitants of the old world.

The Gibbons are the smallest, slenderest, and longest-limbed of the man-like apes : their arms are longer in proportion to their bodies than those of any of the other man-like Apes, so that they can touch the ground when erect ; their hands are longer than their feet, and they are the only Anthropoids which possess callosities like the lower monkeys. They are variously coloured. The Orangs have arms which reach to the ankles in the erect position of the animal ; their thumbs and great toes are very short, and their feet are longer than their hands. They are covered with reddish-brown hair, and the sides of the face, in adult males, are commonly produced into two crescentic, flexible excrescences, like fatty tumours. The Chimpanzees have arms which reach below the knees ; they have large thumbs and great toes, their hands are longer than their feet, and their hair is black, while the skin of the face is pale. The Gorilla, lastly, has arms which reach to the middle of the leg, large thumbs, and great toes, feet longer than the hands, a black face, and dark-grey or dun hair.

For the purpose which I have at present in view, it is unnecessary that I should enter into any further minutiæ respecting the distinctive characters of the genera and species into which these man-like Apes are divided by naturalists. Suffice it to say, that the Orangs and the Gibbons constitute the distinct genera, *Simia* and *Hylobates;* while the Chimpanzees and Gorillas are by some regarded simply as distinct species of one genus, *Troglodytes;* by others as distinct genera—*Troglodytes* being reserved for the Chimpanzees, and *Gorilla* for the Engéena or Pongo.

Sound knowledge respecting the habits and mode of life of the man-like Apes has been even more difficult of attainment than correct information regarding their structure.

. Once in a generation, a Wallace may be found physically, mentally, and morally qualified to wander unscathed through the tropical wilds of America and of Asia; to form magnificent collections as he wanders; and withal to think out sagaciously the conclusions suggested by his collections: but, to the ordinary explorer or collector, the dense forests of equatorial Asia and Africa, which constitute the favourite habitation of the Orang, the Chimpanzee, and the Gorilla, present difficulties of no ordinary magnitude: and the man who risks his life by even a short visit to the malarious shores of those regions may well be excused if he shrinks from facing the dangers of the interior; if he contents himself with stimulating the industry of the better seasoned natives, and collecting and collating the more or less mythical reports and traditions with which they are too ready to supply him.

In such a manner most of the earlier accounts of the habits of the man-like Apes originated; and even now a good deal of what passes current must be admitted to have no very safe foundation. The best information we possess is that, based almost wholly on direct European testimony, respecting the Gibbons ; the next best evidence relates to the Orangs ; while our knowledge of the habits of the Chimpanzee and the Gorilla stands much in need of support and enlargement by additional testimony from instructed European eye-witnesses.

It will therefore be convenient in endeavouring to form a notion of what we are justified in believing about these animals, to commence with the best known man-like Apes, the Gibbons and Orangs ; and to make use of the perfectly reliable information respecting them as a sort of criterion of the probable truth or falsehood of assertions respecting the others.

Of the GIBBONS, half a dozen species are found scattered over the Asiatic islands, Java, Sumatra, Borneo, and through Malacca, Siam, Arracan, and an uncertain extent of Hindostan on the main land of Asia. The largest attain a few inches above three feet in height, from the crown to the heel, so that they are shorter than the other man-like Apes ; while the slenderness of their bodies renders their mass far smaller in proportion even to this diminished height.

Dr. Salomon Müller, an accompl'shed Dutch naturalist, who lived for many years in the Eastern Archipelago, and to the results of whose personal experience I shall frequently have occasion to refer, states that the Gibbons are true mountaineers, loving the slopes and edges of the hills, though they rarely ascend beyond the limit of the fig-trees. All day long they haunt the tops of the tall trees ; and though, towards evening, they descend in small troops to the open ground, no sooner do they spy a man than they dart up the hillsides and disappear in the darker valleys.

All observers testify to the prodigious volume of voice

possessed by these animals.   According to the writer
whom I have just cited, in one of them, the Siamang,

FIG. 8.—A Gibbon (*H. pileatus*), after Wolf.

"the voice is grave and penetrating, resembling the sounds gōek, gōek, gōek, gōek, goek ha ha ha ha haaāāā, and may be easily heard at a distance of half a league." While the cry is being uttered, the great membranous bag under the throat which communicates with the organ of voice, the so-called "laryngeal sac," becomes greatly distended, diminishing again when the creature relapses into silence.

M. Duvaucel, likewise, affirms that the cry of the Siamang may be heard for miles—making the woods ring again. So Mr. Martin* describes the cry of the agile Gibbon as "over-powering and deafening" in a room, and "from its strength, well calculated for resounding through the vast forests." Mr. Waterhouse, an accomplished musician as well as zoologist, says, "The Gibbon's voice is certainly much more powerful than that of any singer I ever heard." And yet it is to be recollected that this animal is not half the height of, and far less bulky in proportion than, a man.

There is good testimony that various species of Gibbon readily take to the erect posture. Mr. George Bennett,† a very excellent observer, in describing the habits of a male *Hylobates syndactylus* which remained for some time in his possession, says : "He invariably walks in the erect posture when on a level surface ; and then the arms either hang down, enabling him to assist himself with his knuckles ; or what is more usual, he keeps his arms uplifted in nearly an erect position, with the hands pendent ready to seize a rope, and climb up on the approach of danger or on the obtrusion of strangers. He walks rather quick in the erect posture, but with a waddling gait, and is soon run down if, whilst pursued, he has no opportunity

* "Man and Monkies," p. 423.
† "Wanderings in New South Wales, Vol. II. chap. viii. 1834.

of escaping by climbing. . . . When he walks in the erect posture, he turns the leg and foot outwards, which occasions him to have a waddling gait and to seem bow-legged."

Dr. Burrough states of another Gibbon, the Horlack or Hooluk :

" They walk erect ; and when placed on the floor, or in an open field, balance themselves very prettily by rais-ing their hands over their head and slightly bending the arm at the wrist and elbow, and then run tolerably fast, rocking from side to side ; and, if urged to greater speed, they let fall their hands to the ground, and assist them-selves forward, rather jumping than running, still keeping the body, however, nearly erect."

Somewhat different evidence, however, is given by Dr. Winslow Lewis :[*]

" Their only manner of walking was on their posterior or inferior extremities, the others being raised upwards to preserve their equilibrium, as rope-dancers are assisted by long poles at fairs. Their progression was not by placing one foot before the other, but by simultaneously using both, as in jumping." Dr. Salomon Müller also states that the Gibbons progress upon the ground by short series of tottering jumps, effected only by the hind limbs, the body being held altogether upright.

But, Mr. Martin, (l. c. p. 418) who also speaks from direct observation, says of the Gibbons generally :

" Pre-eminently qualified for arboreal habits, and dis-playing among the branches amazing activity, the Gib-bons are not so awkward or embarrassed on a level surface as might be imagined. They walk erect, with a waddling or unsteady gait, but at a quick pace ; the equilibrium of

---

[*] Boston Journal of Natural History, Vol. I. 1834.

the body requiring to be kept up, either by touching the
ground with the knuckles, first on one side then on the
other, or by uplifting the arms so as to poise it. As with
the Chimpanzee, the whole of the narrow, long sole of the
foot is placed upon the ground at once and raised at once,
without any elasticity of step."

After this mass of concurrent and independent testimony, it cannot reasonably be doubted that the Gibbons
commonly and habitually assume the erect attitude.

But level ground is not the place where these animals
can display their very remarkable and peculiar locomotive
powers, and that prodigious activity which almost tempts
one to rank them among flying, rather than among ordinary climbing mammals.

Mr. Martin (l. c. p. 430) has given so excellent and
graphic an account of the movements of a *Hylobates
agilis*, living in the Zoological Gardens, in 1840, that I
will quote it in full :

" It is almost impossible to convey in words an idea of
the quickness and graceful address of her movements :
they may indeed be termed aerial as she seems merely to
touch in her progress the branches among which she exhibits her evolutions. In these feats her hands and arms
are the sole organs of locomotion ; her body hanging as if
suspended by a rope, sustained by one hand (the right, for
example), she launches herself, by an energetic movement,
to a distant branch, which she catches with the left hand ;
but her hold is less than momentary : the impulse for the
next launch is acquired : the branch then aimed at is attained by the right hand again, and quitted instantane-
ously, and so on, in alternate succession. In this manner
spaces of twelve and eighteen feet are cleared, with the
greatest ease and uninterruptedly, for hours together,
without the slightest appearance of fatigue being mani

fested ; and it is evident that, if more space could be allowed, distances very greatly exceeding eighteen feet would be as easily cleared ; so that Duvaucel's assertion that he has seen these animals launch themselves from one branch to another, forty feet asunder, startling as it is, may be well credited. Sometimes, on seizing a branch in her progress, she will throw herself by the power of one arm only, completely round it, making a revolution with such rapidity as almost to deceive the eye, and continue her progress with undiminished velocity. It is singular to observe how suddenly this Gibbon can stop, when the impetus given by the rapidity and distance of her swinging leaps would seem to require a gradual abatement of her movements. In the very midst of her flight a branch is seized, the body raised, and she is seen, as if by magic, quietly seated on it, grasping it with her feet. As suddenly she again throws herself into action.

" The following facts will convey some notion of her dexterity and quickness. A live bird was let loose in her apartment ; she marked its flight, made a long swing to a distant branch, caught the bird with one hand in her passage, and attained the branch with her other hand ; her aim, both at the bird and at the branch, being as successful as if one object only had engaged her attention. It may be added that she instantly bit off the head of the bird, picked its feathers, and then threw it down without attempting to eat it.

" On another occasion this animal swung herself from a perch, across a passage at least twelve feet wide, against a window which it was thought would be immediately broken : but not so ; to the surprise of all, she caught the narrow framework between the panes with her hand, in an instant attained the proper impetus, and sprang back again to the cage she had left—a feat requiring not only great strength, but the nicest precision."

The Gibbons appear to be naturally very gentle, but there is very good evidence that they will bite severely when irritated—a female *Hylobates agilis* having so severely lacerated one man with her long canines, that he died; while she had injured others so much that, by way of precaution, these formidable teeth had been filed down; but, if threatened, she would still turn on her keeper. The Gibbons eat insects, but appear generally to avoid animal food. A Siamang, however, was seen by Mr. Bennett to seize and devour greedily a live lizard. They commonly drink by dipping their fingers in the liquid and then licking them. It is asserted that they sleep in a sitting posture.

Duvaucel affirms that he has seen the females carry their young to the waterside and there wash their faces, in spite of resistance and cries. They are gentle and affectionate in captivity—full of tricks and pettishness, like spoiled children, and yet not devoid of a certain conscience, as an anecdote, told by Mr. Bennett (l. c. p. 156), will show. It would appear that his Gibbon had a peculiar inclination for disarranging things in the cabin. Among these articles, a piece of soap would especially attract his notice, and for the removal of this he had been once or twice scolded. "One morning," says Mr. Bennett, "I was writing, the ape being present in the cabin, when casting my eyes towards him, I saw the little fellow taking the soap. I watched him without his perceiving that I did so: and he occasionally would cast a furtive glance towards the place where I sat. I pretended to write; he, seeing me busily occupied, took the soap, and moved away with it in his paw. When he had walked half the length of the cabin, I spoke quietly, without frightening him. The instant he found I saw him, he walked back again, and deposited the soap nearly in the same place from whence he had taken it. There was

FIG. 9.—An adult male Orang-Utan, after Müller and Schlegel.

certainly something more than instinct in that action : he evidently betrayed a consciousness of having done wrong both by his first and last actions—and what is reason if that is not an exercise of it ? "

The most elaborate account of the natural history of the ORANG-UTAN extant, is that given in the " Verhandelingen over de Natuurlijke Geschiedenis · der Nederlandsche overzeesche Bezittingen (1839–45)," by Dr. Salomon Müller and Dr. Schlegel, and I shall base what I have to say upon this subject almost entirely on their statements, adding, here and there, particulars of interest from the writings of Brooke, Wallace, and others.

The Orang-Utan would rarely seem to exceed four feet in height, but the body is very bulky, measuring two-thirds of the height in circumference.*

The Orang-Utan is found only in Sumatra and Borneo, and is common in neither of these islands—in both of which it occurs always in low, flat plains, never in the mountains. It loves the densest and most sombre of the forests, which extend from the sea-shore inland, and thus is found only in the eastern half of Sumatra, where alone

---

* The largest Orang-Utan, cited by Temminck, measured, when standing upright, four feet; but he mentions having just received news of the capture of an Orang five feet three inches high. Schlegel and Müller say that their largest old male measured, upright, 1.25 Netherlands "el;" and from the crown to the end of the toes, 1.5 el ; the circumference of the body being about 1 el. The largest old female was 1.09 el high, when standing. The adult skeleton in the College of Surgeons' Museum, if set upright, would stand 3 feet 0-8 in. from crown to sole. Dr. Humphry gives 3 ft. 8 in. as the mean height of two Orangs. Of seventeen Orangs examined by Mr. Wallace, the largest was 4 ft. 2 in. high, from the heel to the crown of the head. Mr Spencer St. John, however, in his " Life in the Forests of the Far East," tells us of an Orang of " 5 ft. 2 in., measuring fairly from the head to the heel," 15 in. across the face, and 12 in. round the wrist. It does not appear, however, that Mr. St. John measured this Orang himself.

such forests occur, though, occasionally, it strays over to the western side.

On the other hand, it is generally distributed through Borneo, except in the mountains, or where the population is dense. In favourable places, the hunter may, by good fortune, see three or four in a day.

Except in the pairing time, the old males usually live by themselves. The old females, and the immature males, on the other hand, are often met with in twos and threes; and the former occasionally have young with them, though the pregnant females usually separate themselves, and sometimes remain apart after they have given birth to their offspring. The young Orangs seem to remain unusually long under their mother's protection, probably in consequence of their slow growth. While climbing, the mother always carries her young against her bosom, the young holding on by his mother's hair.* At what time of life the Orang-Utan becomes capable of propagation, and how long the females go with young, is unknown, but it is probable that they are not adult until they arrive at ten or fifteen years of age. A female which lived for five years at Batavia, had not attained one-third the height of the wild females. It is probable that, after reaching adult years, they go on growing, though slowly, and that they live to forty or fifty years. The Dyaks tell of old Orangs, which have not only lost all their teeth, but which find it so troublesome to climb, that they maintain themselves on windfalls and juicy herbage.

The Orang is sluggish, exhibiting none of that marvel-

---

* See Mr. Wallace's account of an infant " Orang-Utan," in the " Annals of Natural History" for 1856. Mr. Wallace provided his interesting charge with an artificial mother of buffalo-skin, but the cheat was too successful. The infant's entire experience led it to associate teats with hair, and feeling the latter, it spent its existence in vain endeavours to discover the former.

lous activity characteristic of the Gibbons. Hunger alone seems to stir him to exertion, and when it is stilled, he relapses into repose. When the animal sits, it curves its back and bows its head, so as to look straight down on the ground; sometimes it holds on with its hands by a higher branch, sometimes lets them hang phlegmatically down by its side—and in these positions the Orang will remain, for hours together, in the same spot, almost without stirring, and only now and then giving utterance to its deep, growling voice. By day, he usually climbs from one tree-top to another, and only at night descends to the ground, and if then threatened with danger, he seeks refuge among the underwood. When not hunted, he remains a long time in the same locality, and sometimes stops for many days on the same tree—a firm place among its branches serving him for a bed. It is rare for the Orang to pass the night in the summit of a large tree, probably because it is too windy and cold there for him; but, as soon as night draws on, he descends from the height and seeks out a fit bed in the lower and darker part, or in the leafy top of a small tree, among which he prefers Nibong Palms, Pandani, or one of those parasitic Orchids which give the primæval forests of Borneo so characteristic and striking an appearance. But wherever, he determines to sleep, there he prepares himself a sort of nest: little boughs and leaves are drawn together round the selected spot, and bent crosswise over one another, while to make the bed soft, great leaves of Ferns, of Orchids, of *Pandanus fascicularis*, *Nipa fruticans*, &c., are laid over them. Those which Müller saw, many of them being very fresh, were situated at a height of ten to twenty-five feet above the ground, and had a circumference, on the average, of two or three feet. Some were packed many inches thick with *Pandanus* leaves; others were remarkable only for the cracked twigs, which, united in a common centre, formed a regular platform. "The rude *hut*," says Sir James Brooke, "which they are stated to build in the trees, would be more properly called a seat or nest, for it has no roof or cover of any sort. The facility with which they form this nest is curious, and I had an opportunity of seeing a wounded female weave the branches together and seat herself, within a minute."

According to the Dyaks the Orang rarely leaves his bed before the sun is well above the horizon and has dissipated the mists. He gets up about nine, and goes to bed again about five; but sometimes not till late in the twilight. He lies sometimes on his back; or, by way of change, turns on one side or the other, drawing his limbs up to his body, and resting his head on his hand. When the night is cold, windy, or rainy, he usually covers his body with a heap of *Pandanus*, *Nipa*, or Fern leaves, like those of which his bed is made, and he is especially careful to wrap up his head in them. It is this habit of covering himself up which has probably led to the fable that the Orang builds huts in the trees.

Although the Orang resides mostly amid the boughs of great trees, during the daytime, he is very rarely seen squatting on a thick branch as other apes and particularly the Gibbons, do. The Orang, on the contrary, confines himself to the slender leafy branches, so that he is seen right at the top of the trees, a mode of life which is closely related to the constitution of his hinder limbs, and especially to that of his seat. For this is provided with no callosities, such as are possessed by many of the lower apes, and even by the Gibbons; and those bones of the pelvis, which are termed the ischia, and which form the solid framework of the surface on which the body rests in the sitting posture,

are not expanded like those of the apes which possess callosities, but are more like those of man.

An Orang climbs so slowly and cautiously,[*] as, in this
act, to resemble a man more than an ape, taking great
care of his feet, so that injury of them seems to affect him
far more than it does other apes. Unlike the Gibbons,
whose forearms do the greater part of the work, as they
swing from branch to branch, the Orang never makes
even the smallest jump. In climbing, he moves alter-
nately one hand and one foot, or after having laid fast
hold with the hands, he draws up both feet together. In
passing from one tree to another, he always seeks out a
place where the twigs of both come close together, or in-
terlace. Even when closely pursued, his circumspection
is amazing : he shakes the branches to see if they will bear
him, and then bending an overhanging bough down by
throwing his weight gradually along it, he makes a bridge
from the tree he wishes to quit to the next.[†]

On the ground the Orang always goes laboriously and
shakily, on all fours. At starting he will run faster than
a man, though he may soon be overtaken. The very long
arms which, when he runs, are but little bent, raise the
body of the Orang remarkably, so that he assumes much
the posture of a very old man bent down by age, and
making his way along by the help of a stick. In walking,
the body is usually directed straight forward, unlike the
other apes, which run more or less obliquely ; except the
Gibbons, who in these, as in so many other respects, depart remarkably from their fellows.

---

[*] " They are the slowest and least active of all the monkey tribe, and their
motions are surprisingly awkward and uncouth."—Sir James Brooke, in the
" Proceedings of the Zoological Society," 1841.

[†] Mr. Wallace's account of the progression of the Orang almost exactly
corresponds with this.

3

The Orang cannot put its feet flat on the ground, but is supported upon their outer edges, the heel resting more on the ground, while the curved toes partly rest upon the ground by the upper side of their first joint, the two outermost toes of each foot completely resting on this surface. The hands are held in the opposite manner, their inner edges serving as the chief support. The fingers are then bent out in such a manner that their foremost joints, especially those of the two innermost fingers, rest upon the ground by their upper sides, while the point of the free and straight thumb serves as an aditional fulcrum.

The Orang never stands on its hind legs, and all the pictures, representing it as so doing, are as false as the assertion that it defends itself with sticks, and the like.

The long arms are of especial use, not only in climbing, but in the gathering of food from boughs to which the animal could not trust his weight. Figs, blossoms, and young leaves of various kinds, constitute the chief nutriment of the Orang ; but strips of bamboo two or three feet long were found in the stomach of a male. They are not known to eat living animals.

Although, when taken young, the Orang-Utan soon becomes domesticated, and indeed seems to court human society, it is naturally a very wild and shy animal, though apparently sluggish and melancholy. The Dyaks affirm, that when the old males are wounded with arrows only, they will occasionally leave the trees and rush raging upon their enemies, whose sole safety lies in instant flight, as they are sure to be killed if caught.*

---

* Sir James Brooke, in a letter to Mr. Waterhouse, published in the proceedings of the Zoological Society for 1841, says:—" On the habits of the Orangs, as far as I have been able to observe them, I may remark that they are as dull and slothful as can well be conceived, and on no occasion, when pursuing them, did they move so fast as to preclude my keeping pace with

But, though possessed of immense strength, it is rare for the Orang to attempt to defend itself, especially when attacked with fire-arms. On such occasions he endeavours to hide himself, or to escape along the topmost branches of the trees, breaking off and throwing down the boughs as he goes. When wounded he betakes himself to the highest attainable point of the tree, and emits a singular cry, consisting at first of high notes, which at length deepen into a low roar, not unlike that of a panther. While giving out the high notes the Orang thrusts out his lips into a funnel shape; but in uttering the low notes he holds his mouth wide open, and at the same time the great throat bag, or laryngeal sac, becomes distended.

According to the Dyaks, the only animal the Orang measures his strength with is the crocodile, who occasionally seizes him on his visits to the water side. But they

them easily through a moderately clear forest; and even when obstructions below (such as wading up to the neck) allowed them to get away some distance, they were sure to stop and allow me to come up. I never observed the slightest attempt at defence, and the wood which sometimes rattled about our ears was broken by their weight, and not thrown, as some persons represent. If pushed to extremity, however, the *Pappan* could not be otherwise than formidable, and one unfortunate man, who, with a party, was trying to catch a large one alive, lost two of his fingers, besides being severely bitten on the face, whilst the animal finally beat off his pursuers and escaped."

Mr. Wallace, on the other hand, affirms that he has several times observed them throwing down branches when pursued. "It is true he does not throw them *at* a person, but casts them down vertically; for it is evident that a bough cannot be thrown to any distance from the top of a lofty tree. In one case a female Mias, on a durian tree, kept up for at least ten minutes a continuous shower of branches and of the heavy, spined fruits, as large as 32-pounders, which most effectually kept us clear of the tree she was on. She could be seen breaking them off and throwing them down with every appearance of rage, uttering at intervals a loud pumping grunt, and evidently meaning mischief."—" On the habits of the Orang-Utan," Annals of Nat. History. 1856. This statement, it will be observed, is quite in accordance with that contained in the letter of the Resident Palm quoted above (p. 10).

say that the Orang is more than a match for his enemy, and beats him to death, or rips up his throat by pulling the jaws asunder !

Much of what has been here stated was probably derived by Dr. Müller from the reports of his Dyak hunters; but a large male, four feet high, lived in captivity under his observation, for a month, and receives a very bad character.

"He was a very wild beast," says Müller, "of prodigious strength, and false and wicked to the last degree. If any one approached he rose up slowly with a low growl, fixed his eyes in the direction in which he meant to make his attack, slowly passed his hand between the bars of his cage, and then extending his long arm, gave a sudden grip —usually at the face." He never tried to bite (though Orangs will bite one another), his great weapons of offence and defence being his hands.

His intelligence was very great; and Müller remarks, that, though the faculties of the Orang have been estimated too highly, yet Cuvier, had he seen this specimen, would not have considered its intelligence to be only a little higher than that of a dog.

His hearing was very acute, but the sense of vision seemed to be less perfect. The under lip was the great organ of touch, and played a very important part in drinking, being thrust out like a trough, so as either to catch the falling rain, or to receive the contents of the half cocoa-nut shell full of water with which the Orang was supplied, and which, in drinking, he poured into the trough thus formed.

In Borneo the Orang-Utan of the Malays goes by the name of "*Mias*" among the Dyaks, who distinguish several kinds as *Mias Pappan*, or *Zimo*, *Mias Kassu*, and *Mias Rambi*. Whether these are distinct species, how-

ever, or whether they are mere races, and how far any of
them are identical with the Sumatran Orang, as Mr. Wallace thinks the Mias Pappan to be, are problems which
are at present undecided; and the variability of these
great apes is so extensive, that the settlement of the question is a matter of great difficulty. Of the form called
" Mias Pappan," Mr. Wallace* observes, " It is known
by its large size, and by the lateral expansion of the face
into fatty protuberances or ridges, over the temporal muscles, which have been mis-termed *callosities*, as they are
perfectly soft, smooth, and flexible. Five of this form,
measured by me, varied only from 4 feet 1 inch to 4 feet
2 inches in height, from the heel to the crown of the head,
the girth of the body from 3 feet to 3 feet 7½ inches, and
the extent of the outstretched arms from 7 feet 2 inches to
7 feet 6 inches; the width of the face from 10 to 13¼
inches. The colour and length of the hair varied in different individuals, and in different parts of the same individual; some possessed a rudimentary nail on the great toe,
others none at all; but they otherwise present no external
differences on which to establish even varieties of a
species.

Yet, when we examine the crania of these individuals,
we find remarkable differences of form, proportion, and
dimension, no two being exactly alike. The slope of the
profile, and the projection of the muzzle, together with
the size of the cranium, offer differences as decided as
those existing between the most strongly marked forms of
the Caucasian and African crania in the human species.
The orbits vary in width and height, the cranial ridge is
either single or double, either much or little developed,
and the zygomatic aperture varies considerably in size.

* On the Orang-Utan, or Mias of Borneo, Annals of Natural History,
1856.

This variation in the proportions of the crania enables us satisfactorily to explain the marked difference presented by the single-crested and double-crested skulls, which have been thought to prove the existence of two large species of Orang. · The external surface of the skull varies considerably in size, as do also the zygomatic aperture and the temporal muscle ; but they bear no necessary relation to each other, a small muscle often existing with a large cranial surface, and *vice versâ*. Now, those skulls which have the largest and strongest jaws and the widest zygomatic aperture, have the muscles so large that they meet on the crown of the skull, and deposit the bony ridge which separates them, and which is the highest in that which has the smallest cranial surface. In those which combine a large surface with comparatively weak jaws, and small zygomatic aperture, the muscles, on each side, do not extend to the crown, a space of from 1 to 2 inches remaining between them, and along their margins small ridges are formed. Intermediate forms are found, in which the ridges meet only in the hinder part of the skull. The form and size of the ridges are therefore independent of age, being sometimes more strongly developed in the less aged animal. Professor Temminck states that the series of skulls in the Leyden Museum shows the same result."

Mr. Wallace observed two male adult Orangs (Mias Kassu of the Dyaks), however, so very different from any of these that he concludes them to be specially distinct ; they were respectively 3 feet 8½ in. and 3 feet 9½ inches high, and possessed no sign of the cheek excrescences, but otherwise resembled the larger kinds. The skull has no crest, but two bony ridges, 1¾ inches to 2 inches apart, as in the *Simia morio* of Professor Owen. The teeth, however, are immense, equalling or surpassing those of the other species. The females of both these kinds, according to Mr. Wallace, are devoid of excrescences, and resemble the smaller males, but are shorter by 1½ to 3 inches, and their canine teeth are comparatively small, subtruncated and dilated at the base, as in the so-called *Simia morio*, which is, in all probability, the skull of a female of the same species as the smaller males. Both males and females of this smaller species are distinguishable, according to Mr. Wallace, by the comparatively large size of the middle incisors of the upper jaw.

So far as I am aware, no one has attempted to dispute the accuracy of the statements which I have just quoted regarding the habits of the two Asiatic man-like Apes; and if true, they must be admitted as evidence, that such an Ape—

1stly, May readily move along the ground in the erect, or semi-erect, position, and without direct support from its arms.

2ndly, That it may possess an extremely loud voice, so loud as to be readily heard one or two miles.

3rdly, That it may be capable of great viciousness and violence when irritated: and this is especially true of adult males.

4thly, That it may build a nest to sleep in.

Such being well-established facts respecting the Asiatic Anthropoids, analogy alone might justify us in expecting the African species to offer similar peculiarities, separately or combined; or, at any rate, would destroy the force of any attempted *à priori* argument against such direct testimony as might be adduced in favour of their existence. And, if the organization of any of the African Apes could be demonstrated to fit it better than either of its Asiatic allies for the erect position and for efficient attack, there would be still less reason for doubting its occasional adoption of the upright attitude or of aggressive proceedings.

From the time of Tyson and Tulpius downwards, the habits of the young CHIMPANZEE in a state of captivity have been abundantly reported and commented upon. But trustworthy evidence as to the manners and customs of adult anthropoids of this species, in their native woods, was almost wanting up to the time of the publication of the paper by Dr. Savage, to which I have already referred ; containing notes of the observations which he made, and of the information which he collected from sources which he considered trustworthy, while resident at Cape Palmas, at the northwestern limit of the Bight of Benin.

The adult Chimpanzees, measured by Dr. Savage, never exceeded, though the males may almost attain, five feet in height.

" When at rest, the sitting posture is that generally assumed. They are sometimes seen standing and walking, but when thus detected, they immediately take to all fours, and flee from the presence of the observer. Such is their organization that they cannot stand erect, but lean forward. Hence they are seen, when standing, with the hands clasped over the occiput, or the lumbar region, which would seem necessary to balance or ease of posture.

" The toes of the adult are strongly flexed and turned inwards, and cannot be perfectly straightened. In the attempt the skin gathers into thick folds on the back, shewing that the full expansion of the foot, as is necessary in walking, is unnatural. The natural position is on all fours, the body anteriorly resting upon the knuckles. These are greatly enlarged, with the skin protuberant and thickened like the sole of the foot.

" They are expert climbers, as one would suppose from their organization. In their gambols · they swing from limb to limb at a great distance, and leap with astonishing agility. It is not unusual to see the ' old folks' (in the language of an observer) sitting under a tree, regaling themselves with fruit and friendly chat, while their ' children' are leaping around them, and swinging from tree to tree with boisterous merriment.

" As seen here, they cannot be called *gregarious*, seldom more than five, or ten at most, being found together. It has been said, on good authority, that they occasionally assemble in large numbers, in gambols. My informant asserts that he saw once not less than fifty so engaged ; hooting, screaming, and drumming with sticks upon old logs, which is done in the latter case with equal facility by the four extremities. They do not appear ever to act on the offensive, and seldom, if ever really, on the defensive. When about to be captured, they resist by throwing their arms about their opponent, and attempting to draw him into contact with their teeth." (Savage, l. c. p. 384.)

With respect to this last point Dr. Savage is very explicit in another place :

" *Biting* is their principal art of defence. I have seen one man who had been thus severely wounded in the feet.

" The strong development of the canine teeth in the adult would seem to indicate a carnivorous propensity ; but in no state save that of domestication do they manifest it. At first they reject flesh, but easily acquire a fondness for it. The canines are early developed, and evidently designed to act the important part of weapons of defence. When in contact with man almost the first effort of the animal is—*to bite*.

" They avoid the abodes of men, and build their habitations in trees. Their construction is more that of *nests* than *huts*, as they have been erroneously termed by some naturalists. They generally build not far above the ground. Branches or twigs are bent, or partly broken, and crossed, and the whole supported by the body of a limb or a crotch. Sometimes a nest will be found near the *end* of a *strong leafy branch* twenty or thirty feet from the ground. One I have lately seen that could not be less than forty feet, and more probably it was fifty. But this is an unusual height.

" Their dwelling-place is not permanent, but changed in pursuit of food and solitude, according to the force of circumstances. We more often see them in elevated places ; but this arises from the fact that the low grounds, being more favourable for the natives' rice-farms, are the oftener cleared, and hence are almost always wanting in suitable trees for their nests. . . . It is seldom that more than one or two nests are seen upon the same tree, or in the same neighbourhood : five have been found, but it was an unusual circumstance." . . . . . .

" They are very filthy in their habits. . . . . It is a tradition with the natives generally here, that they were once members of their own tribe : that for their depraved habits they were expelled from all human society, and that through an obstinate indulgence of their vile propensities, they have degenerated into their present state of organization. They are, however, eaten by them, and when cooked with the oil and pulp of the palm-nut considered a highly palatable morsel.

" They exhibit a remarkable degree of intelligence in their habits, and, on the part of the mother, much affection for their young. The second female described was upon a tree when first discovered, with her mate and two young ones (a male and a female). Her first impulse was to descend with great rapidity, and make off into the thicket, with her mate and female offspring. The young male remaining behind, she soon returned to the rescue. She ascended and took him in her arms, at which moment she was shot, the ball passing through the fore-arm of the young one, on its way to the heart of the mother. . . . .

"In a recent case, the mother, when discovered, remained upon the tree with her offspring, watching intently the movements of the hunter. As he took aim, she motioned with her hand, precisely in the manner of a human being, to have him desist and go away. When the wound has not proved instantly fatal, they have been known to stop the flow of blood by pressing with the hand upon the part, and when they did not succeed, to apply leaves and grass. . . . . When shot, they give a sudden screech, not unlike that of a human being in sudden and acute distress."

"The ordinary voice of the Chimpanzee, however, is affirmed to be hoarse, guttural, and not very loud, somewhat like ' whoo-whoo.' " (l. c. p. 365.)

The analogy of the Chimpanzee to the Orang, in its nest-building habit and in the mode of forming its nest, is exceedingly interesting ; while, on the other hand, the activity of this ape, and its tendency to bite, are particulars in which it rather resembles the Gibbons. In extent of geographical range, again, the Chimpanzees—which are found from Sierra Leone to Congo—remind one of the Gibbons, rather than of either of the other man-like apes ; and it seems not unlikely that, as is the case with the Gibbons, there may be several species spread over the geographical area of the genus.

The same excellent observer, from whom I have borrowed the preceding account of the habits of the adult Chimpanzee, published, fifteen years ago,[*] an account of

* Notice of the external characters and habits of Troglodytes Gorilla. Boston Journal of Natural History, 1847.

the GORILLA, which has, in its most essential points, been confirmed by subsequent observers, and to which so very little has really been added, that in justice to Dr. Savage I give it almost in full.

FIG. 10.—The Gorilla, after Wolf.

" It should be borne in mind that my account is based upon the statements of the aborigines of that region (the Gaboon). In this connection, it may also be proper for me to remark, that having been a missionary resident for several years, studying, from habitual intercourse, the African mind and character, I felt myself prepared to discriminate and decide upon the probability of their statements. Besides, being familiar with the history and habits of its interesting congener (*Trog. niger*, Geoff.), I was able to separate their accounts of the two animals, which, having the same locality and a similarity of habit, are confounded in the minds of the mass, especially as but few— such as traders to the interior and huntsmen—have ever seen the animal in question.

The tribe from which our knowledge of the animal is derived, and whose territory forms its habitat, is the *Mpongwe*, occupying both banks of the River Gaboon, from its mouth to some fifty or sixty miles upward. . . .

If the word " Pongo " be of African origin, it is probably a corruption of the word *Mpongwe*, the name of the tribe on the banks of the Gaboon, and hence applied to the region they inhabit. Their local name for the Chimpanzee is *Enché-eko*, as near as it can be Anglicized, from which the common term " Jocko " probably comes. The Mpongwe appellation for its new congener is *Engé-ena*, prolonging the sound of the first vowel, and slightly sounding the second.

The habitat of the *Engé-ena* is the interior of lower Guinea, whilst that of the *Enché-eko* is nearer the seaboard.

Its height is about five feet; it is disproportionately broad across the shoulders, thickly covered with coarse black hair, which is said to be similar in its arrangement to that of the *Enché-eko;* with age it becomes gray, which fact has given rise to the report that both animals are seen of different colours.

Head.—The prominent features of the head are, the great width and elongation of the face, the depth of the molar region, the branches of the lower jaw being very deep and extending far backward, and the comparative smallness of the cranial portion ; the eyes are very large, and said to be like those of the Enché-cko, a bright hazel ; nose broad and flat, slightly elevated towards the root ; the muzzle broad, and prominent lips and chin, with scattered gray hairs ; the under lip highly mobile, and capable of great elongation when the animal is enraged, then hanging over the chin ; skin of the face and ears naked, and of a dark brown, approaching to black.

The most remarkable feature of the head is a high ridge, or crest of hair, in the course of the sagittal suture, which meets posteriorly with a transverse ridge of the same, but less prominent, running round from the back of one ear to the other. The animal has the power of moving the scalp freely forward and back, and when enraged is said to contract it strongly over the brow, thus bringing down the hairy ridge and pointing the hair forward, so as to present an indescribably ferocious aspect.

Neck short, thick, and hairy ; chest and shoulders very broad, said to be fully double the size of the Enché-ekos ; arms very long, reaching some way below the knee—the fore-arm much the shortest ; hands very large, the thumbs much larger than the fingers. . . . . .

The gait is shuffling ; the motion of the body, which is never upright as in man, but bent forward, is somewhat rolling, or from side to side. The arms being longer than the Chimpanzee, it does not stoop as much in walking ; like that animal, it makes progression by thrusting its arms forward, resting the hands on the ground, and then giving the body a half jumping half swinging motion between them. In this act it is said not to flex the fingers, as does the Chimpanzee, resting on its knuckles, but to extend them, making a fulcrum of the hand. When it assumes the walking posture, to which it is said to be much inclined, it balances its huge body by flexing its arms upward.

Fig. 11.—Gorilla walking (after Wolff.)

They live in bands, but are not so numerous as the Chimpanzees: the females generally exceed the other sex in number. My informants all agree in the assertion that but one adult male is seen in a band; that when the young males grow up, a contest takes place for mastery, and the strongest, by killing and driving out the others, establishes himself as the head of the community."

Dr. Savage repudiates the stories about the Gorillas carrying off women and vanquishing elephants, and then adds—

"Their dwellings, if they may be so called, are similar to those of the Chimpanzee, consisting simply of a few sticks and leafy branches, supported by the crotches and limbs of trees: they afford no shelter, and are occupied only at night.

"They are exceedingly ferocious, and always offensive in their habits, never running from man, as does the Chimpanzee. They are objects of terror to the natives, and are never encountered by them except on the defensive. The few that have been captured were killed by elephant-hunters and native traders, as they came suddenly upon them while passing through the forests.

" It is said, that when the male is first seen he gives a terrific yell, that resounds far and wide through the forest, something like kh—ah! kh—ah! prolonged and shrill. His enormous jaws are widely opened at each expiration, his under lip hangs over the chin, and the hairy ridge and scalp are contracted upon the brow, presenting an aspect of indescribable ferocity.

" The females and young, at the first cry, quickly disappear. He then approaches the enemy in great fury, pouring out his horrid cries in quick succession. The hunter awaits his approach with his gun extended : if his aim is not sure, he permits the animal to grasp the barrel, and as he carries it to his mouth (which is his habit) he fires. Should the gun fail to go off, the barrel (that of the ordinary musket, which is thin) is crushed between his teeth, and the encounter soon proves fatal to the hunter.

" In the wild state, their habits are in general like those of the *Troglodytes niger*, building their nests loosely in trees, living on similar fruits, and changing their place of resort from force of circumstances."

Dr. Savage's observations were confirmed and supplemented by those of Mr. Ford, who communicated an interesting paper on the Gorilla to the Philadelphian Academy of Sciences, in 1852. With respect to the geographical distribution of this greatest of all the man-like Apes, Mr. Ford remarks :

" This animal inhabits the range of mountains that traverse the interior of Guinea, from the Cameroon in the north, to Angola in the south, and about 100 miles inland, and called by the geographers Crystal Mountains. The limit to which this animal extends, either north or south, I am unable to define. But that limit is doubtless some

distance north of this river [Gaboon]. I was able to cer-
tify myself of this fact in a late excursion to the head-
waters of the Mooney (Danger) River, which comes into
the sea some sixty miles from this place. I was informed
(credibly, I think,) that they were numerous among the
mountains in which that river rises, and far north of that.

" In the south, this species extends to the Congo River,
as I am told by native traders who have visited the coast,
between the Gaboon and that river. Beyond that, I am
not informed. This animal is only found at a distance
from the coast in most cases, and, according to my best
information, approaches it nowhere so nearly as on the
south side of this river, where they have been found within
ten miles of the sea. This, however, is only of late occurrence. I am informed by some of the oldest Mpongwe
men that formerly he was only found on the sources of the
river, but that at present he may be found within half-a-
day's walk of its mouth. Formerly he inhabited the
mountainous ridge where Bushmen alone inhabited, but
now he boldly approaches the Mpongwe plantations.
This is doubtless the reason of the scarcity of information
in years past, as the opportunities for receiving a knowledge of the animal have not been wanting ; traders having
for one hundred years frequented this river, and specimens, such as have been brought here within a year, could
not have been exhibited without having attracted the attention of the most stupid."

One specimen Mr. Ford examined weighed 170lbs.,
without the thoracic, or pelvic, viscera, and measured four
feet four inches round the chest. This writer describes so
minutely and graphically the onslaught of the Gorilla—
though he does not for a moment pretend to have wit-
nessed the scene—that I am tempted to give this part of
his paper in full, for comparison with other narratives :

" He always rises to his feet when making an attack, though he approaches his antagonist in a stooping posture.

" Though he never lies in wait, yet, when he hears, sees, or scents a man, he immediately utters his characteristic cry, prepares for an attack, and always acts on the offensive. The cry he utters resembles a grunt more than a growl, and is similar to the cry of the Chimpanzee, when irritated, but vastly louder. It is said to be audible at a great distance. His preparation consists in attending the females and young ones, by whom he is usually accompanied, to a little distance. He, however, soon returns, with his crest erect and projecting forward, his nostrils dilated, and his under-lip thrown down ; at the same time uttering his characteristic yell, designed, it would seem, to terrify his antagonist. Instantly, unless he is disabled by a well-directed shot, he makes an onset, and, striking his antagonist with the palm of his hands, or seizing him with a grasp from which there is no escape, he dashes him upon the ground, and lacerates him with his tusks.

" He is said to seize a musket, and instantly crush the barrel between his teeth. . . . . . This animal's savage nature is very well shewn by the implacable desperation of a young one that was brought here. It was taken very young, and kept four months, and many means were used to tame it ; but it was incorrigible, so that it bit me an hour before it died."

Mr. Ford discredits the house-building and elephant-driving stories, and says that no well-informed natives believe them. They are tales told to children.

I might quote other testimony to a similar effect, but, as it appears to me, less carefully weighed and sifted, from the letters of MM. Franquet and Gautier Laboullay, appended to the memoir of M. I. G. St. Hilaire, which I have already cited.

Bearing in mind what is known regarding the Orang and the Gibbon, the statements of Dr. Savage and Mr. Ford do not appear to me to be justly open to criticism on *à priori* grounds. The Gibbons, as we have seen, readily assume the erect posture, but the Gorilla is far better fitted by its organization for that attitude than are the Gibbons : if the laryngeal pouches of the Gibbons, as is very likely, are important in giving volume to a voice which can be heard for half a league, the Gorilla, which has similar sacs, more largely developed, and whose bulk is fivefold that of a Gibbon, may well be audible for twice that distance. If the Orang fights with its hands, the Gibbons and Chimpanzees with their teeth, the Gorilla may, probably enough, do either or both ; nor is there anything to be said against either Chimpanzee or Gorilla building a nest, when it is proved that the Orang-Utan habitually performs that feat.

With all this evidence, now ten to fifteen years old, before the world, it is not a little surprising that the assertions of a recent traveller, who, so far as the Gorilla is concerned, really does very little more than repeat, on his own authority, the statements of Savage and Ford, should have met with so much and such bitter opposition. If subtraction be made of what was known before, the sum and substance of what M. Du Chaillu has affirmed as a matter of his own observation respecting the Gorilla, is, that, on advancing to the attack, the great brute beats his chest with his fists. I confess I see nothing very improbable, or very much worth disputing about, in this statement.

With respect to the other man-like Apes of Africa, M. Du Chaillu tells us absolutely nothing, of his own knowledge, regarding the common Chimpanzee ; but he informs us of a bald-headed species or variety, the *nschiego mbouve*, which builds itself a shelter, and of another rare kind with a comparatively small face, large facial angle, and peculiar note, resembling " Kooloo."

As the Orang shelters itself with a rough coverlet of leaves, and the common Chimpanzee, according to that eminently trustworthy observer Dr. Savage, makes a sound like " Whoo-whoo,"—the grounds of the summary repudiation with which M. Du Chaillu's statements on these matters have been met is not obvious.

If I have abstained from quoting M. Du Chaillu's work, then, it is not because I discern any inherent improbability in his assertions respecting the man-like Apes ; nor from any wish to throw suspicion on his veracity ; but because, in my opinion, so long as his narrative remains in its present state of unexplained and apparently inexplicable confusion, it has no claim to original authority respecting any subject whatsoever.

It may be truth, but it is not evidence.

AFRICAN CANNIBALISM IN THE SIXTEENTH CENTURY.

In turning over Pigafetta's version of the narrative of Lopez, which I have quoted above, I came upon so curious and unexpected an anticipation, by some two centuries and a half, of one of the most startling parts of M. Du Chaillu's narrative, that I cannot refrain from drawing attention to it in a note, although I must confess that the subject is not strictly relevant to the matter in hand.

In the fifth chapter of the first book of the " Descriptio, " Concerning the northern part of the Kingdom of Congo and its boundaries, is mentioned a people whose king is called ' Maniloango,' and who live under the equator, and as far westward as Cape Lopez. This appears to be the country now inhabited by the Ogobai and Bakalai according to M. Du Chaillu. — " Beyond these dwell another people called ' Anziques,' of incredible ferocity, for they eat one another, sparing neither friends nor relations."

Fig. 12.—Butcher's Shop of the Anziques, Anno 1598.

These people are armed with small bows bound tightly round with snake skins, and strung with a reed or rush. Their arrows, short and slender, but made of hard wood, are shot with great rapidity. They have iron axes, the handles of which are bound with snake skins, and swords with scabbards of the same material; for defensive armour they employ elephant hides. They cut their skins when young, so as to produce scars. "Their butchers' shops are filled with human flesh instead of that of oxen or sheep. For they eat the enemies whom they take in battle. They fatten, slay, and devour their slaves also, unless they think they shall get a good price for them; and, moreover, sometimes for weariness of life or desire for glory (for they think it a great thing and the sign of a generous soul to despise life), or for love of their rulers, offer themselves up for food."

"There are indeed many cannibals, as in the Eastern Indies and in Brazil and elsewhere, but none such as these, since the others only eat their enemies, but these their own blood relations."

The careful illustrators of Pigafetta have done their best to enable the reader to realize this account of the 'Anziques,' and the unexampled butcher's shop represented in fig. 12, is a facsimile of part of their Plate XII.

M. Du Chaillu's account of the Fans accords most singularly with what Lopez here narrates of the Anziques. He speaks of their small crossbows and little arrows, of their axes and knives, "ingeniously sheathed in snake skins." "They tattoo themselves more than any other tribes I have met with north of the equator." And all the world knows what M. Du Chaillu says of their cannibalism—"Presently we passed a woman who solved all doubt. She bore with her a piece of the thigh of a human body, just as we should go to market and carry thence a roast or steak." M. Du Chaillu's artist cannot generally be accused of any want of courage in embodying the statements of his author, and it is to be regretted that, with so good an excuse, he has not furnished us with a fitting companion to the sketch of the brothers De Bry.

II.

ON THE RELATIONS OF MAN TO THE LOWER ANIMALS.

Multis videri poterit, majorem esse differentiam Simiæ et Hominis, quam diei et noctis; verum tamen hi, comparatione instituta inter summos Europæ Heroës et Hottentottos ad Caput bonæ spei degentes, difficillime sibi persuadebunt, has eosdem habere natales; vel si virginem nobilem aulicam, maxime contam et humanissimam, conferre vellent cum homine sylvestri et sibi relicto, vix augurari possent, hunc et illam ejusdem esse speciei.—Linnæi Amœnitates Acad. "Anthropomorpha."

THE question of questions for mankind—the problem which underlies all others, and is more deeply interesting than any other—is the ascertainment of the place which Man occupies in nature and of his relations to the universe of things. Whence our race has come; what are the limits of our power over nature, and of nature's power over us; to what goal we are tending; are the problems which present themselves anew and with undiminished interest to every man born into the world. Most of us, shrinking from the difficulties and dangers which beset the seeker after original answers to these riddles, are contented to ignore them altogether, or to smother the investigating spirit under the featherbed of respected and respectable tradition. But, in every age, one or two restless spirits, blessed with that constructive genius, which can only build on a secure foundation, or cursed with the mere spirit of scepticism, are unable to follow in the well-worn and comfortable track of their forefathers and contemporaries, and unmindful of thorns and stumbling-blocks, strike out into paths of their own. The sceptics end in the infidelity which asserts the problem to be insoluble, or in the atheism which denies the existence of any orderly progress and governance of things: the men of genius propound solutions which grow into systems of Theology or of Philosophy, or veiled in musical language which suggests more than it asserts, take the shape of the Poetry of an epoch.

Each such answer to the great question, invariably asserted by the followers of its propounder, if not by himself, to be complete and final, remains in high authority and esteem, it may be for one century, or it may be for twenty: but, as invariably, Time proves each reply to have been a mere approximation to the truth—tolerable chiefly on account of the ignorance of those by whom it was accepted, and wholly intolerable when tested by the larger knowledge of their successors.

In a well-worn metaphor, a parallel is drawn between the life of man and the metamorphosis of the caterpillar into the butterfly; but the comparison may be more just as well as more novel, if for its former term we take the mental progress of the race. History shows that the human mind, fed by constant accessions of knowledge, periodically grows too large for its theoretical coverings, and bursts them asunder to appear in new habiliments, as the feeding and growing grub, at intervals, casts its too narrow skin and assumes another, itself but temporary. Truly the imago state of Man seems to be terribly distant, but every moult is a step gained, and of such there have been many.

Since the revival of learning, whereby the Western

races of Europe were enabled to enter upon that progress towards true knowledge, which was commenced by the philosophers of Greece, but was almost arrested in subsequent long ages of intellectual stagnation, or, at most, gyration, the human larva has been feeding vigorously, and moulting in proportion. A skin of some dimension was cast in the 16th century, and another towards the end of the 18th, while, within the last fifty years, the extraordinary growth of every department of physical science has spread among us mental food of so nutritious and stimulating a character that a new ecdysis seems imminent. But this is a process not unusually accompanied by many throes and some sickness and debility, or, it may be, by graver disturbances ; so that every good citizen must feel bound to facilitate the process, and even if he have nothing but a scalpel to work withal, to ease the cracking integument to the best of his ability.

In this duty lies my excuse for the publication of these essays. For it will be admitted that some knowledge of man's position in the animate world is an indispensable preliminary to the proper understanding of his relations to the universe—and this again resolves itself, in the long run, into an inquiry into the nature and the closeness of the ties which connect him with those singular creatures whose history* has been sketched in the preceding pages.

The importance of such an inquiry is indeed intuitively manifest. Brought face to face with these blurred copies of himself, the least thoughtful of men is conscious of a certain shock, due, perhaps, not so much to disgust at the aspect of what looks like an insulting caricature, as to the awakening of a sudden and profound mistrust of time-

* It will be understood that, in the preceding Essay, I have selected for notice from the vast mass of papers which have been written upon the man-like Apes, only those which seem to me to be of special moment.

4

' honoured theories and strongly-rooted prejudices regarding his own position in nature, and his relations to the under-world of life; while that which remains a dim suspicion for the unthinking, becomes a vast argument, fraught with the deepest consequences, for all who are acquainted with the recent progress of the anatomical and physiological sciences.

I now propose briefly to unfold that argument, and to set forth, in a form intelligible to those who possess no special acquaintance with anatomical science, the chief facts upon which all conclusions respecting the nature and the extent of the bonds which connect man with the brute world must be based: I shall then indicate the one immediate conclusion which, in my judgment, is justified by those facts, and I shall finally discuss the bearing of that conclusion upon the hypotheses which have been entertained respecting the Origin of Man.

The facts to which I would first direct the reader's attention, though ignored by many of the professed instructors of the public mind, are easy of demonstration and are universally agreed to by men of science; while their significance is so great, that whoso has duly pondered over them will, I think, find little to startle him in the other revelations of Biology. I refer to those facts which have been made known by the study of Development.

It is a truth of very wide, if not of universal, application, that every living creature commences its existence under a form different from, and simpler than, that which it eventually attains.

The oak is a more complex thing than the little rudimentary plant contained in the acorn; the caterpillar is more complex than the egg; the butterfly than the caterpillar; and each of these beings, in passing from its rudimentary to its perfect condition, runs through a series of changes, the sum of which is called its Development. In the higher animals these changes are extremely complicated; but, within the last half century, the labours of such men as Von Baer, Rathke, Reichert, Bischof, and Remak, have almost completely unravelled them, so that the successive stages of development which are exhibited by a Dog, for example, are now as well known to the embryologist as are the steps of the metamorphosis of the silk-worm moth to the school-boy. It will be useful to consider with attention the nature and the order of the stages of canine development, as an example of the process in the higher animals generally.

The Dog, like all animals, save the very lowest (and further inquiries may not improbably remove the apparent exception), commences its existence as an egg: as a body which is, in every sense, as much an egg as that of a hen, but is devoid of that accumulation of nutritive matter which confers upon the bird's egg its exceptional size and domestic utility; and wants the shell, which would not only be useless to an animal incubated within the body of its parent, but would cut it off from access to the source of that nutriment which the young creature requires, but which the minute egg of the mammal does not contain within itself.

The Dog's egg is, in fact, a little spheroidal bag (Fig. 13), formed of a delicate transparent membrane called the *vitelline membrane*, and about $\frac{1}{130}$ to $\frac{1}{120}$th of an inch in diameter. It contains a mass of viscid nutritive matter— the '*yelk*'—within which is inclosed a second much more delicate spheroidal bag, called the '*germinal vesicle*' (*a*). In this, lastly, lies a more solid rounded body, termed the '*germinal spot*' (*b*).

The egg, or 'Ovum,' is originally formed within a gland, from which, in due season, it becomes detached,

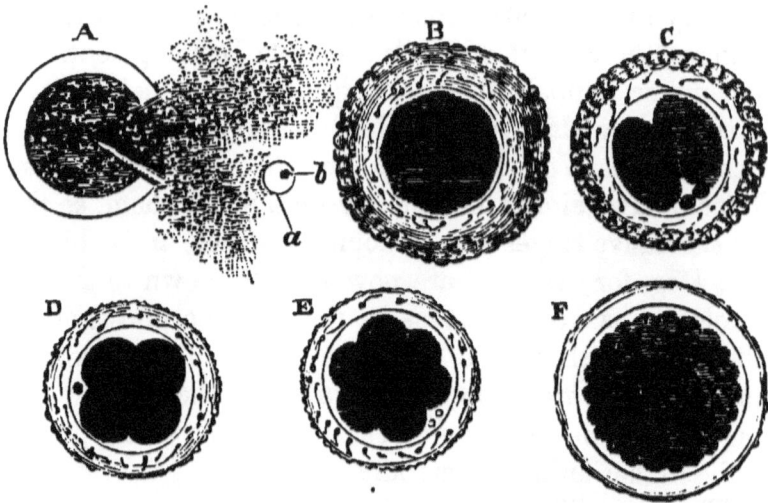

Fig. 13.—A. Egg of the Dog, with the vitelline membrane burst, so as to give exit to the yelk, the germinal vesicle (*a*), and its included spot (*b*).

B. C. D. E. F. Successive changes of the yelk indicated in the text. After Bischoff.

and passes into the living chamber fitted for its protection and maintenance during the protracted process of gestation. Here, when subjected to the required conditions, this minute and apparently insignificant particle of living matter, becomes animated by a new and mysterious activity. The germinal vesicle and spot cease to be discernible (their precise fate being one of the yet unsolved problems of embryology), but the yelk becomes circumferentially indented, as if an invisible knife had been drawn round it, and thus appears divided into two hemispheres (Fig. 13, C). .

By the repetition of this process in various planes,

these hemispheres become subdivided, so that four segments are produced (D) ; and these, in like manner, divide
and subdivide again, until the whole yelk is converted
into a mass of granules, each of which consists of a minute
spheroid of yelk-substance, inclosing a central particle, the
so-called '*nucleus*' (F). Nature, by this process, has attained much the same result as that at which a human
artificer arrives by his operations in a brick field. She
takes the rough plastic material of the yelk and breaks it
up into well-shaped tolerably even-sized masses—handy
for building up into any part of the living edifice.

Next, the mass of organic bricks, or '*cells*' as they are
technically called, thus formed, acquires an orderly arrangement, becoming converted into a hollow spheroid
with double walls. Then, upon one side of this spheroid,
appears a thickening, and, by and bye, in the centre of
the area of thickening, a straight shallow groove (Fig. 14,
A) marks the central line of the edifice which is to be
raised, or, in other words, indicates the position of the
middle line of the body of the future dog. The substance
bounding the groove on each side next rises up into a fold,
the rudiment of the side wall of that long cavity, which
will eventually lodge the spinal marrow and the brain ;
and in the floor of this chamber appears a solid cellular
cord, the so-called '*notochord*.' One end of the inclosed
cavity dilates to form the head (Fig. 14, B), the other remains narrow, and eventually becomes the tail ; the side
walls of the body are fashioned out of the downward continuation of the walls of the groove ; and from them, by
and bye, grow out little buds which, by degrees, assume
the shape of limbs. Watching the fashioning process stage
by stage, one is forcibly reminded of the modeller in clay.
Every part, every organ, is at first, as it were, pinched up
rudely, and sketched out in the rough ; then shaped more

accurately, and only, at last, receives the touches which stamp its final character.

Thus, at length, the young puppy assumes such a form as is shewn in Fig. 14, C. In this condition it has a dis-

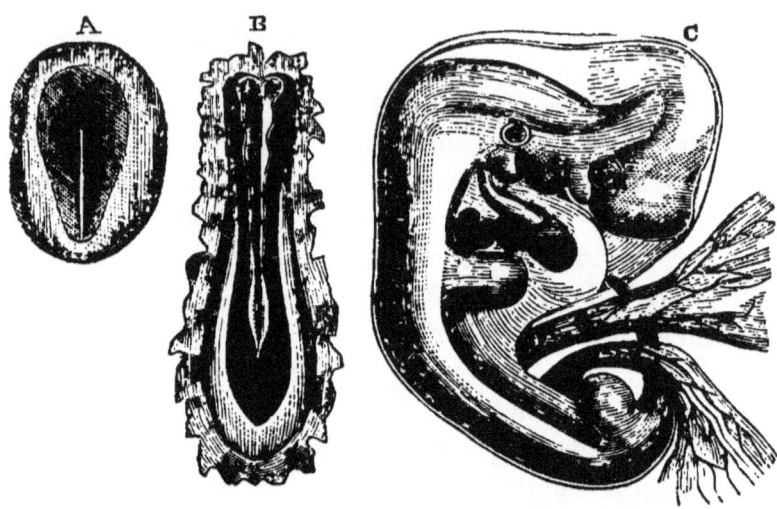

Fig. 14.—A. Earliest rudiment of the Dog. B. Rudiment further advanced, showing the foundations of the head, tail, and vertebral column. C. The very young puppy, with attached ends of the yelk-sac and allantois, and invested in the amnion.

proportionately large head, as dissimilar to that of a dog as the bud-like limbs are unlike his legs.

The remains of the yelk, which have not yet been applied to the nutrition and growth of the young animal, are contained in a sac attached to the rudimentary intestine, and termed the yelk sac, or ' *umbilical vesicle.*' Two membranous bags, intended to subserve respectively the protection and nutrition of the young creature, have been developed from the skin and from the under and hinder surface of the body ; the former, the so-called ' *amnion,*' is a sac filled with fluid, which invests the whole body of the embryo, and plays the part of a sort of water-bed for it; the other, termed the '*allantois*,' grows out, loaded with blood-vessels, from the ventral region, and eventually applying itself to the walls of the cavity, in which the developing organism is contained, enables these vessels to become the channel by which the stream of nutriment, required to supply the wants of the offspring, is furnished to it by the parent.

The structure which is developed by the interlacement of the vessels of the offspring with those of the parent, and by means of which the former is enabled to receive nourishment and to get rid of effete matters, is termed the '*Placenta*.'

It would be tedious, and it is unnecessary for my present purpose, to trace the process of development further; suffice it to say, that, by a long and gradual series of changes, the rudiment here depicted and described, becomes a puppy, is born, and then, by still slower and less perceptible steps, passes into the adult Dog.

There is not much apparent resemblance between a barn-door Fowl and the Dog who protects the farm-yard. Nevertheless the student of development finds, not only that the chick commences its existence as an egg, primarily identical, in all essential respects, with that of the Dog, but that the yelk of this egg undergoes division—that the primitive groove arises, and that the contiguous parts of the germ are fashioned, by precisely similar methods, into a young chick, which, at one stage of its existence, is so like the nascent Dog, that ordinary inspection would hardly distinguish the two.

The history of the development of any other vertebrate animal, Lizard, Snake, Frog, or Fish, tells the same story. There is always, to begin with, an egg having the same essential structure as that of the Dog :—the yelk of that egg always undergoes division, or '*segmentation*' as it is often called : the ultimate products of that segmentation constitute the building materials for the body of the young animal; and this is built up round a primitive groove, in the floor of which a notochord is developed. Furthermore there is a period in which the young of all these animals resemble one another, not merely in outward form, but in all essentials of structure, so closely, that the differences between them are inconsiderable, while, in their subsequent course, they diverge more and more widely from one another. And it is a general law, that, the more closely any animals resemble one another in adult structure, the longer and the more intimately do their embryos resemble one another; so that, for example, the embryos of a Snake and of a Lizard remain like one another longer than do those of a Snake and of a Bird ; and the embryo of a Dog and of a Cat remain like one another for a far longer period than do those of a Dog and a Bird ; or of a Dog and an Opossum ; or even than those of a Dog and a Monkey.

Thus the study of development affords a clear test of closeness of structural affinity, and one turns with impatience to inquire what results are yielded by the study of the development of Man. Is he something apart ? Does he originate in a totally different way from Dog, Bird, Frog, and Fish, thus justifying those who assert him to have no place in nature and no real affinity with the lower world of animal life ? Or does he originate in a similar germ, pass through the same slow and gradually progressive modifications,—depend on the same contrivances for protection and nutrition, and finally enter the world by the help of the same mechanism? The reply is not doubtful for a moment, and has not been doubtful any time these thirty years. Without question, the mode of origin and the early stages of the development of man are identical with those of the animals immediately below him in the scale :—without a doubt, in these respects, he is far nearer the Apes, than the Apes are to the Dog.

The Human ovum is about $\frac{1}{125}$ of an inch in diameter, and might be described in the same terms as that of the Dog, so that I need only refer to the figure illustrative (15 A.) of its structure. It leaves the organ in which it is formed in a similar fashion and enters the organic chamber prepared for its reception in the same way, the conditions of its development being in all respects the same. It has not yet been possible (and only by some rare chance can it ever be possible) to study the human ovum in so early a developmental stage as that of yelk division, but there is every reason to conclude that the changes it undergoes are identical with those exhibited by the ova of other vertebrated animals ; for the formative materials of which the rudimentary human body is composed, in the earliest conditions in which it has been observed, are the same as those of other animals. Some of these earliest stages are figured below and, as will be seen, they are strictly comparable to the very early states of the Dog ; the marvellous correspondence between the two which is kept up, even for some time, as development advances, becoming apparent by the simple comparison of the figures with those on page 63.

Indeed, it is very long before the body of the young human being can be readily discriminated from that of the young puppy ; but, at a tolerably early period, the two become distinguishable by the different form of their adjuncts, the yelk-sac and the allantois. The former, in the Dog, becomes long and spindle-shaped, while in Man it remains spherical : the latter, in the Dog, attains an ex-

4*

tremely large size, and the vascular processes which are developed from it and eventually give rise to the formation of the placenta (taking root, as it were, in the parental

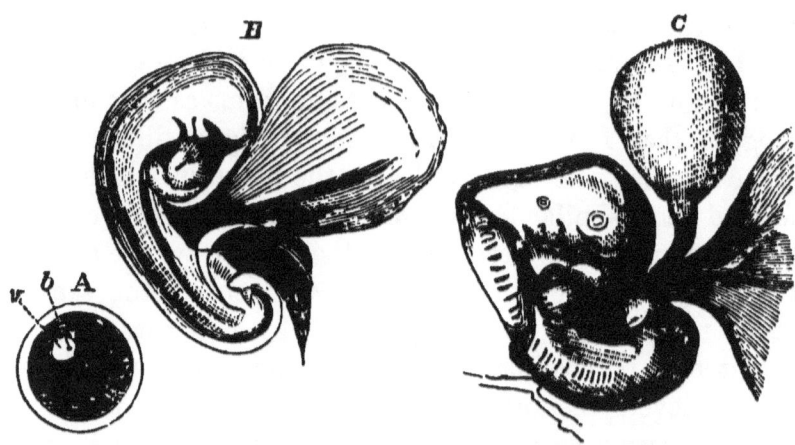

Fig. 15.—A. Human ovum (after Kölliker). *a.* germinal vesicle. *b.* germinal spot.

B. A very early condition of Man, with yelk-sac, allantois and amnion (original).

C. A more advanced stage (after Kölliker), compare fig. 14, C.

organism, so as to draw nourishment therefrom, as the root of a tree extracts it from the soil) are arranged in an encircling zone, while in Man, the allantois remains comparatively small, and its vascular rootlets are eventually restricted to one disk-like spot. Hence, while the placenta of the Dog is like a girdle, that of Man has the cake-like form, indicated by the name of the organ.

But, exactly in those respects in which the developing Man differs from the Dog, he resembles the Ape, which, like man, has a spheroidal yelk-sac and a discoidal—sometimes partially lobed-placenta.

So that it is only quite in the later stages of development that the young human being presents marked differences from the young ape, while the latter departs as, much from the dog in its development as the man does.

Startling as the last assertion may appear to be, it is demonstrably true, and it alone appears to me sufficient to place beyond all doubt the structural unity of man with the rest of the animal world, and more particularly and closely with the apes.

Thus, identical in the physical processes by which he originates—identical in the early stages of his formation—identical in the mode of his nutrition before and after birth, with the animals which lie immediately below him in the scale—Man, if his adult and perfect structure be compared with theirs, exhibits, as might be expected, a marvellous likeness of organization. He resembles them as they resemble one another—he differs from them as they differ from one another.—And, though these differences and resemblances cannot be weighed and measured, their value may be readily estimated ; the scale or standard of judgment, touching that value, being afforded and expressed by the system of classification of animals now current among zoologists.

A careful study of the resemblances and differences presented by animals has, in fact, led naturalists to arrange them into groups, or assemblages, all the members of each group presenting a certain amount of definable resemblance, and the number of points of similarity being smaller as the group is larger and *vicé versâ*. Thus, all creatures which agree only in presenting the few distinctive marks of animality form the 'Kingdom' ANIMALIA. The numerous animals which agree only in possessing the special characters of Vertebrates form one 'Sub-kingdom' of this Kingdom. Then the Sub-kingdom VERTEBRATA is subdivided into the five 'Classes,' Fishes, Amphibians,

Reptiles, Birds, and Mammals, and these into smaller groups called 'Orders;' these into 'Families' and 'Genera;' while the last are finally broken up into the smallest assemblages, which are distinguished by the possession of constant, not-sexual, characters. These ultimate groups are Species.

Every year tends to bring about a greater uniformity of opinion throughout the zoological world as to the limits and characters of these groups, great and small. At present, for example, no one has the least doubt regarding the characters of the classes Mammalia, Aves, or Reptilia; nor does the question arise whether any thoroughly well-known animal should be placed in one class or the other. Again, there is a very general agreement respecting the characters and limits of the orders of Mammals, and as to the animals which are structurally necessitated to take a place in one or another order.

No one doubts, for example, that the Sloth and the Ant-eater, the Kangaroo and the Opossum, the Tiger and the Badger, the Tapir and the Rhinoceros, are respectively members of the same orders. These successive pairs of animals may, and some do, differ from one another immensely, in such matters as the proportions and structure of their limbs; the number of their dorsal and lumbar vertebræ; the adaptation of their frames to climbing, leaping, or running; the number and form of their teeth; and the characters of their skulls and of the contained brain. But, with all these differences, they are so closely connected in all the more important and fundamental characters of their organization, and so distinctly separated by these same characters from other animals, that zoologists find it necessary to group them together as members of one order. And if any new animal were discovered, and were found to present no greater difference.

from the Kangaroo and the Opossum, for example, than these animals do from one another, the zoologist would not only be logically compelled to rank it in the same order with these, but he would not think of doing otherwise.

Bearing this obvious course of zoological reasoning in mind, let us endeavour for a moment to disconnect our thinking selves from the mask of humanity; let us imagine ourselves scientific Saturnians, if you will, fairly acquainted with such animals as now inhabit the Earth, and employed in discussing the relations they bear to a new and singular 'erect and featherless biped,' which some enterprising traveller, overcoming the difficulties of space and gravitation, has brought from that distant planet for our inspection, well preserved, may be, in a cask of rum. We should all, at once, agree upon placing him among the mammalian vertebrates; and his lower jaw, his molars, and his brain, would leave no room for doubting the systematic position of the new genus among those mammals, whose young are nourished during gestation by means of a placenta, or what are called the 'placental mammals.'

Further, the most superficial study would at once convince us that, among the orders of placental mammals, neither the Whales nor the hoofed creatures, nor the Sloths and Ant-eaters, nor the carnivorous Cats, Dogs, and Bears, still less the Rodent Rats and Rabbits, or the Insectivorous Moles and Hedgehogs, or the Bats, could claim our '*Homo*' as one of themselves.

There would remain then, but one order for comparison, that of the Apes (using that word in its broadest sense), and the question for discussion would narrow itself to this—is Man so different from any of these Apes that he must form an order by himself? Or does he differ

less from them than they differ from one another, and
hence must take his place in the same order with them?

Being happily free from all real, or imaginary, personal interest in the results of the inquiry thus set afoot,
we should proceed to weigh the arguments on one side and
on the other, with as much judicial calmness as if the
question related to a new Opossum. We should endeavour to ascertain, without seeking either to magnify or
diminish them, all the characters by which our new Mammal differed from the Apes; and if we found that these
were of less structural value, than those which distinguish
certain members of the Ape order from others universally
admitted to be of the same order, we should undoubtedly
place the newly discovered tellurian genus with them.

I now proceed to detail the facts which seem to me to
leave us no choice but to adopt the last mentioned course.

It is quite certain that the Ape which most nearly approaches man, in the totality of its organization, is either
the Chimpanzee or the Gorilla; and as it makes no practical difference, for the purposes of my present argument,
which is selected for comparison, on the one hand, with
Man, and on the other hand, with the rest of the Primates,* I shall select the latter (so far as its organization
is known)—as a brute now so celebrated in prose and
verse, that all must have heard of him, and have formed
some conception of his appearance. I shall take up as
many of the most important points of difference between
man and this remarkable creature, as the space at my dis-
posal will allow me to discuss, and the necessities of the
argument demand; and I shall inquire into the value and

---

* We are not at present thoroughly acquainted with the brain of the Go
rilla, and therefore, in discussing cerebral characters, I shall take that of the
Chimpanzee as my highest term among the Apes.

magnitude of these differences, when placed side by side with those which separate the Gorilla from other animals of the same order.

In the general proportions of the body and limbs there is a remarkable difference between the Gorilla and Man, which at once strikes the eye. The Gorilla's brain-case is smaller, its trunk larger, its lower limbs shorter, its upper limbs longer in proportion than those of Man.

I find that the vertebral column of a full grown Gorilla, in the Museum of the Royal College of Surgeons, measures 27 inches along its anterior curvature, from the upper edge of the atlas, or first vertebra of the neck, to the lower extremity of the sacrum ; that the arm, without the hand, is $31\frac{1}{2}$ inches long ; that the leg, without the foot, is $26\frac{1}{2}$ inches long ; that the hand is $9\frac{3}{4}$ inches long ; the foot $11\frac{1}{4}$ inches long.

In other words, taking the length of the spinal column as 100, the arm equals 115, the leg 96, the hand 36, and the foot 41.

In the skeleton of a male Bosjesman, in the same collection, the proportions, by the same measurement, to the spinal column, taken as 100, are—the arm 78, the leg 110, the hand 26, and the foot 32. In a woman of the same race the arm is 83, and the leg 120, the hand and foot remaining the same. In a European skeleton I find the arm to be 80, the leg 117, the hand 26, the foot 35.

Thus the leg is not so different as it looks at first sight, in its proportions to the spine in the Gorilla and in the Man—being very slightly shorter than the spine in the former, and between $\frac{1}{10}$ and $\frac{1}{5}$ longer than the spine in the latter. The foot is longer and the hand much longer in the Gorilla ; but the great difference is caused by the arms, which are very much longer than the spine in the Gorilla, very much shorter than the spine in the Man.

The question now arises how are the other Apes related to the Gorilla in these respects—taking the length of the spine, measured in the same way, at 100. In an adult Chimpanzee, the arm is only 96, the leg 90, the hand 43, the foot 39—so that the hand and the leg depart more from the human proportion and the arm less, while the foot is about the same as in the Gorilla.

In the Orang, the arms are very much longer than in the Gorilla (122), while the legs are shorter (88); the foot is longer than the hand (52 and 48), and both are much longer in proportion to the spine.

In the other man-like Apes again, the Gibbons, these proportions are still further altered; the length of the arms being to that of the spinal column as 19 to 11; while the legs are also a third longer than the spinal column, so as to be longer than in Man, instead of shorter. The hand is half as long as the spinal column, and the foot, shorter than the hand, is about $\frac{4}{11}$ths of the length of the spinal column.

Thus *Hylobates* is as much longer in the arms than the Gorilla, as the Gorilla is longer in the arms than Man; while, on the other hand, it is as much longer in the legs than the Man, as the Man is longer in the legs than the Gorilla, so that it contains within itself the extremest deviations from the average length of both pairs of limbs (see the Frontispiece).

The Mandrill presents a middle condition, the arms and legs being nearly equal in length, and both being shorter than the spinal column; while hand and foot have nearly the same proportions to one another and to the spine, as in Man.

In the Spider monkey (*Ateles*) the leg is longer than the spine, and the arm than the leg; and, finally, in that remarkable Lemurine form, the Indri, (*Lichanotus*) the leg is about as long as the spinal column, while the arm is not more than $\frac{11}{14}$ of its length; the hand having rather less and the foot rather more, than one third the length of the spinal column.

These examples might be greatly multiplied, but they suffice to show that, in whatever proportion of its limbs the Gorilla differs from Man, the other Apes depart still more widely from the Gorilla and that, consequently, such differences of proportion can have no ordinal value.

We may next consider the differences presented by the trunk, consisting of the vertebral column, or backbone, and the ribs and pelvis, or bony hip-basin, which are connected with it, in Man and in the Gorilla respectively.

In Man, in consequence partly of the disposition of the articular surfaces of the vertebræ, and largely of the elastic tension of some of the fibrous bands, or ligaments, which connect these vertebræ together, the spinal column, as a whole, has an elegant S-like curvature, being convex forwards in the neck, concave in the back, convex in the loins, or lumbar region, and concave again in the sacral region; an arrangement which gives much elasticity to the whole backbone, and diminishes the jar communicated to the spine, and through it to the head, by locomotion in the erect position.

Furthermore, under ordinary circumstances, Man has seven vertebræ in his neck, which are called *cervical;* twelve succeed these, bearing ribs and forming the upper part of the back, whence they are termed *dorsal;* five lie in the loins, bearing no distinct, or free, ribs, and are called *lumbar;* five, united together into a great bone, excavated in front, solidly wedged in between the hip bones, to form the back of the pelvis, and known by the name of the *sacrum,* succeed these; and finally, three or four little more or less moveable bones, so small as to be insignificant, constitute the *coccyx* or rudimentary tail.

In the Gorilla, the vertebral column is similarly divided into cervical, dorsal, lumbar, sacral and coccygeal vertebræ, and the total number of cervical and dorsal vertebræ, taken together, is the same as in man ; but the development of a pair of ribs to the first lumbar vertebra, which is an exceptional occurrence in Man, is the rule in the Gorilla ; and hence, as lumbar are distinguished from dorsal vertebræ only by the presence or absence of free ribs, the seventeen " dorso-lumbar " vertebræ of the Gorilla are divided into thirteen dorsal and four lumbar, while in Man they are twelve dorsal and five lumbar.

Not only, however, does Man occasionally possess thirteen pair of ribs,* but the Gorilla sometimes has fourteen pairs, while an Orang-Utan skeleton in the Museum of the Royal College of Surgeons has twelve dorsal and five lumbar vertebræ, as in Man. Cuvier notes the same number in a *Hylobates*. On the other hand, among the lower Apes, many possess twelve dorsal and six or seven lumbar vertebræ ; the Douroucouli has fourteen dorsal and eight lumbar, and a Lemur (*Stenops tardigradus*) has fifteen dorsal and nine lumbar vertebræ.

The vertebral column of the Gorilla, as a whole, differs from that of Man in the less marked character of its curves, especially in the slighter convexity of the lumbar region. Nevertheless, the curves are present, and are quite obvious in young skeletons of the Gorilla and Chim-

---

* " More than once," says Peter Camper, " have I met with more than six lumbar vertebræ in man. . . . Once I found thirteen ribs and four lumbar vertebræ." Fallopius noted thirteen pair of ribs and only four lumbar vertebræ ; and Eustachius once found eleven dorsal vertebræ and six lumbar vertebræ.—' Œuvres de Pierre Camper,' T. 1, p. 42. As Tyson states, his ' Pygmie ' had thirteen pair of ribs and five lumbar vertebræ. The question of the curves of the spinal column in the Apes requires further investigation.

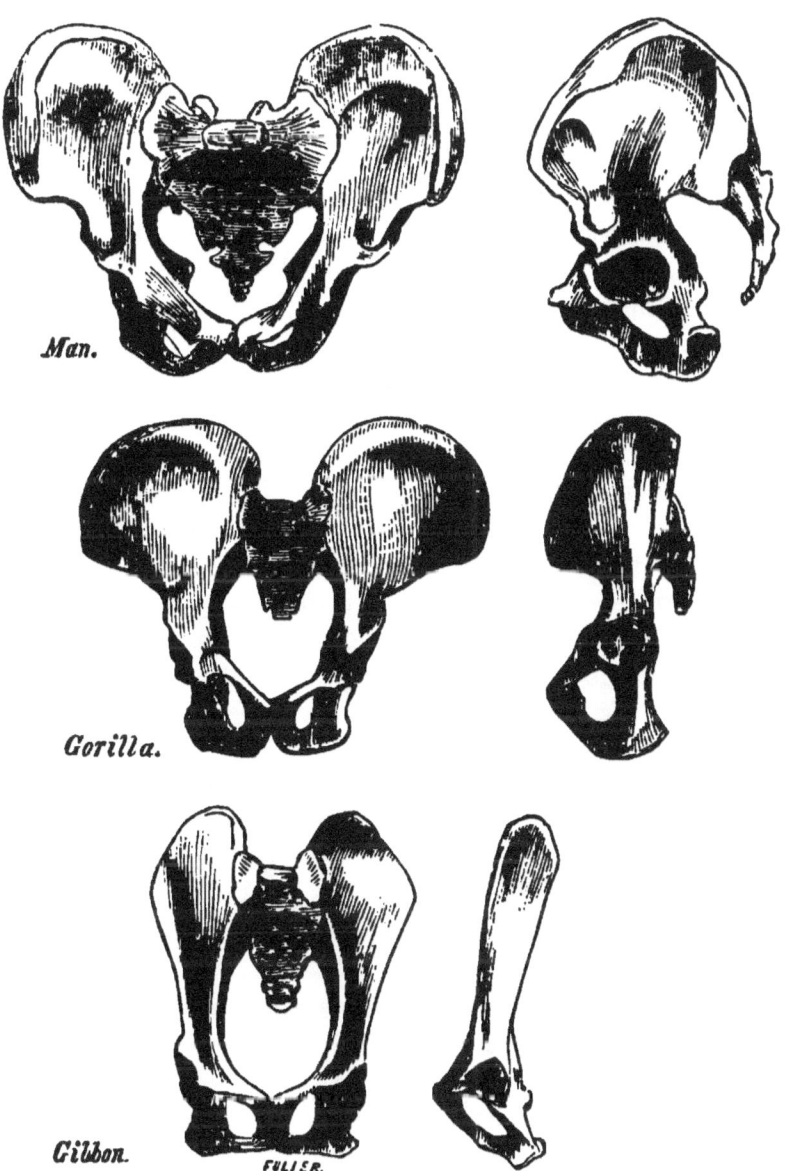

FIG. 16.—Front and side views of the bony pelvis of Man, the Gorilla and Gibbon: reduced from drawings made from nature, of the same absolute length, by Mr. Waterhouse Hawkins.

panzee which have been prepared without removal of the ligaments. In young Orangs similarly preserved, on the other hand, the spinal column is either straight, or even concave forwards, throughout the lumbar region.

Whether we take these characters then, or such minor ones as those which are derivable from the proportional length of the spines in the cervical vertebræ, and the like, there is no doubt whatsoever as to the marked difference between Man and the Gorilla; but there is as little, that equally marked differences, of the very same order, obtain between the Gorilla and the lower apes.

The Pelvis, or bony girdle of the hips, of Man is a strikingly human part of his organization; the expanded haunch bones affording support for his viscera during his habitually erect posture, and giving space for the attachment of the great muscles which enable him to assume and to preserve that attitude. In these respects the pelvis of the Gorilla differs very considerably from his (Fig. 16). But go no lower than the Gibbon, and see how vastly more he differs from the Gorilla than the latter does from Man, even in this structure. Look at the flat, narrow haunch bones—the long and narrow passage—the coarse, outwardly curved, ischiatic prominences on which the Gibbon habitually rests, and which are coated by the so-called "callosities," dense patches of skin, wholly absent in the Gorilla, in the Chimpanzee, and in the Orang, as in Man!

In the lower Monkeys and in the Lemurs the difference becomes more striking still, the pelvis acquiring an altogether quadrupedal character.

But now let us turn to a nobler and more characteristic organ—that by which the human frame seems to be, and indeed is, so strongly distinguished from all others,— I mean the skull. The differences between a Gorilla's skull and a Man's are truly immense (Fig. 17). In the former, the face, formed largely by the massive jaw-bones, predominates over the brain case, or cranium proper : in the latter, the proportions of the two are reversed. In the Man, the occipital foramen, through which passes the great nervous cord connecting the brain with the nerves of the body, is placed just behind the centre of the base of the skull, which thus becomes evenly balanced in the erect posture ; in the Gorilla it lies in the posterior third of that base. In the Man, the surface of the skull is comparatively smooth, and the supraciliary ridges or brow prominences usually project but little—while, in the Gorilla, vast crests are developed upon the skull, and the brow ridges overhang the cavernous orbits, like great penthouses.

Sections of the skulls, however, show that some of the apparent defects of the Gorilla's cranium arise, in fact, not so much from deficiency of brain case as from excessive development of the parts of the face. The cranial cavity is not ill-shaped, and the forehead is not truly flattened or very retreating, its really well-formed curve being simply disguised by the mass of bone which is built up against it (Fig. 17).

But the roofs of the orbits rise more obliquely into the cranial cavity, thus diminishing the space for the lower part of the anterior lobes of the brain, and the absolute capacity of the cranium is far less than that of Man. So far as I am aware, no human cranium belonging to an adult man has yet been observed with a less cubical capacity than 62 cubic inches, the smallest cranium observed in any race of men by Morton, measuring 63 cubic inches ; while, on the other hand, the most capacious Gorilla skull yet measured has a content of not more than $34\frac{1}{2}$ cubic inches. Let us assume, for simplicity's sake, that the lowest Man's skull has twice the capacity of the highest Gorilla.*

No doubt, this is a very striking difference, but it loses much of its apparent systematic value, when viewed by the light of certain other equally indubitable facts respecting cranial capacities.

The first of these is, that the difference in the volume of the cranial cavity of different races of mankind is far greater, absolutely, than that between the lowest Man and the highest Ape, while, relatively, it is about the same.   For the largest human skull measured by Morton, contained 114 cubic inches, that is to say, had very nearly double the capacity of the smallest; while its absolute

---

* It has been affirmed that Hindoo crania sometimes contain as little as 27 ounces of water, which would give a capacity of about 46 cubic inches. The minimum capacity which I have assumed above, however, is based upon the valuable tables published by Professor R. Wagner in his "Vrostudien zu einer wissenschaftlichen Morphologie und Physiologie des menschlichen Gehirns.'   As the result of the careful weighing of more than 900 human brains, Professor Wagner states that one-half weighed between 1200 and 1400 grammes, and that about two-ninths, consisting for the most part of male brains, exceed 1400 grammes.   The lightest brain of an adult male, with sound mental faculties, recorded by Wagner, weighed 1020 grammes.   As a gramme equals 15.4 grains, and a cubic inch of water contains 252.4 grains, this is equivalent to 62 cubic inches of water; so that as brain is heavier than water, we are perfectly safe against erring on the side of diminution in taking this as the smallest capacity of any adult male human brain.   The only adult male brain, weighing as little as 970 grammes, is that of an idiot; but the brain of an adult woman, against the soundness of whose faculties nothing appears, weighed as little as 907 grammes (55.3 cubic inches of water); and Reid gives an adult female brain of still smaller capacity.   The heaviest brain (1872 grammes, or about 115 cubic inches) was, however, that of a woman; next to it comes the brain of Cuvier (1861 grammes), then Byron (1807 grammes), and then an insane person (1783 grammes).   The lightest adult brain recorded (720 grammes) was that of an idiotic female.   The brains of five children, four years old, weighed between 1275 and 992 grammes.   So that it may be safely said, that an average European child of four years old has a brain twice as large as that of an adult Gorilla.

preponderance, of 52 cubic inches—is far greater than that by which the lowest adult male human cranium surpasses the largest of the Gorillas ($62 - 34\frac{1}{2} = 27\frac{1}{2}$). Secondly, the adult crania of Gorillas which have as yet been measured differ among themselves by nearly one-third, the maximum capacity being 34.5 cubic inches, the minimum 24 cubic inches ; and, thirdly, after making all due allowance for difference of size, the cranial capacities of some of the lower apes fall nearly as much, relatively, below those of the higher Apes as the latter fall below Man.

Thus, even in the important matter of cranial capacity, Men differ more widely from one another than they do from the Apes ; while the lowest Apes differ as much, in proportion, from the highest, as the latter does from Man. The last proposition is still better illustrated by the study of the modifications which other parts of the cranium undergo in the Simian series.

It is the large proportional size of the facial bones and the great projection of the jaws which confers upon the Gorilla's skull its small facial angle and brutal character.

But if we consider the proportional size of the facial bones to the skull proper only, the little *Chrysothrix* (Fig. 17) differs very widely from the Gorilla, and in the same way as Man does ; while the Baboons (*Cynocephalus*, Fig. 17) exaggerate the gross proportions of the muzzle of the great Anthropoid, so that its visage looks mild and human by comparison with theirs. The difference between the Gorilla and the Baboon is even greater than it appears at first sight ; for the great facial mass of the former is largely due to a downward development of the jaws ; an essentially human character, superadded upon that almost purely forward, essentially brutal, development of the same parts which characterizes the Baboon, and yet more remarkably distinguishes the Lemur.

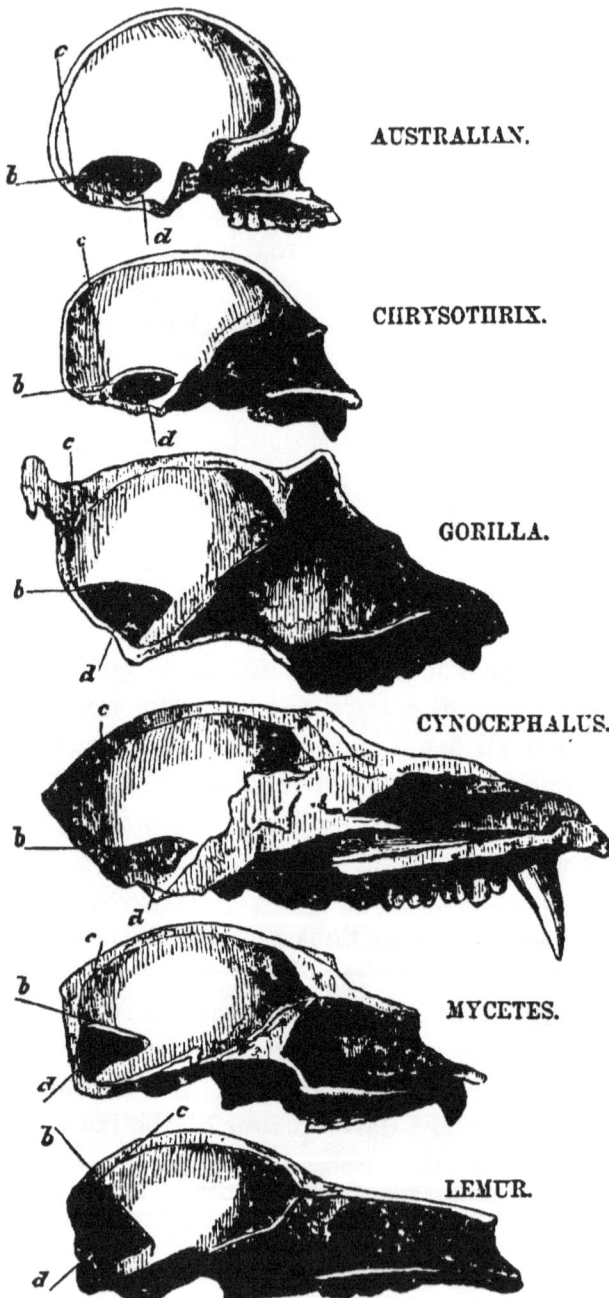

AUSTRALIAN.

CHRYSOTHRIX.

GORILLA.

CYNOCEPHALUS.

MYCETES.

LEMUR.

FIG. 17.—Sections of the skulls of Man and various Apes, drawn so as to give the cerebral cavity the same length in each case, thereby displaying the varying proportions of the facial bones. The line *b* indicates the plane of the tentorium, which separates the cerebrum from the cerebellum; *d*, the axis of the occipital outlet of the skull. The extent of the cerebral cavity behind *c*, which is a perpendicular erected on *b* at the point where the tentorium is attached posteriorly, indicates the degree to which the cerebrum overlaps the cerebellum—the space occupied by which is roughly indicated by the dark shading. In comparing these diagrams, it must be recollected, that figures on so small a scale as these simply exemplify the statements in the text, the proof of which is to be found in the objects themselves.

Similarly the occipital foramen of *Mycetes* (Fig. 17), and still more of the Lemurs, is situated completely in the posterior face of the skull, or as much further back than that of the Gorilla, as that of the Gorilla is further back than that of Man; while, as if to render patent the futility of the attempt to base any broad classificatory distinction on such a character, the same group of Platyrhine, or American monkeys, to which the *Mycetes* belongs, contains the *Chrysothrix*, whose occipital foramen is situated far more forward than in any other ape, and nearly approaches the position it holds in Man.

Again, the Orang's skull is as devoid of excessively developed supraciliary prominences as a Man's, though some varieties exhibit great crests elsewhere (see p. 54); and in some of the Cebine apes and in the *Chrysothrix*, the cranium is as smooth and rounded as that of Man himself.

What is true of these leading characteristics of the skull, holds good, as may be imagined, of all minor features; so that for every constant difference between the Gorilla's skull and the Man's, a similar constant difference of the same order (that is to say, consisting in excess or defect of the same quality) may be found between the Gorilla's skull and that of some other ape. So that, for the skull, no less than for the skeleton in general, the proposition holds good, that the differences between Man and the Gorilla are of smaller value than those between the Gorilla and some other Apes.

In connection with the skull, I may speak of the teeth —organs which have a peculiar classificatory value, and whose resemblances and differences of number, form, and succession, taken as a whole, are usually regarded as more trustworthy indicators of affinity than any others.

5

Man is provided with two sets of teeth—milk teeth and permanent teeth. The former consist of four incisors, or cutting-teeth ; two canines, or eye-teeth ; and four molars, or grinders, in each jaw, making twenty in all. The latter (Fig. 18) comprise four incisors, two canines, four small grinders, called premolars or false molars, and six large grinders, or true molars in each jaw—making thirty-two in all. The internal incisors are larger than the external pair, in the upper jaw, smaller than the external pair, in the lower jaw. The crowns of the upper molars exhibit four cusps, or blunt-pointed elevations, and a ridge crosses the crown obliquely, from the inner, anterior, cusp to the outer, posterior cusp (Fig. 18 m^2). The anterior lower molars have five cusps, three external and two internal. The premolars have two cusps, one internal and one external, of which the outer is the higher.

In all these respects the dentition of the Gorilla may be described in the same terms as that of Man ; but in other matters it exhibits many and important differences (Fig. 18).

Thus the teeth of man constitute a regular and even series—without any break and without any marked projection of one tooth above the level of the rest ; a peculiarity which, as Cuvier long ago showed, is shared by no other mammal save one—as different a creature from man as can well be imagined—namely, the long extinct *Anoplotherium*. The teeth of the Gorilla, on the contrary, exhibit a break, or interval, termed the *diastema*, in both jaws : in front of the eye-tooth, or between it and the outer incisor, in the upper jaw ; behind the eye-tooth, or between it and the front false molar in the lower jaw. Into this break in the series, in each jaw, fits the canine of the opposite jaw ; the size of the eye-tooth in the Gorilla being so great that it projects, like a tusk, far beyond

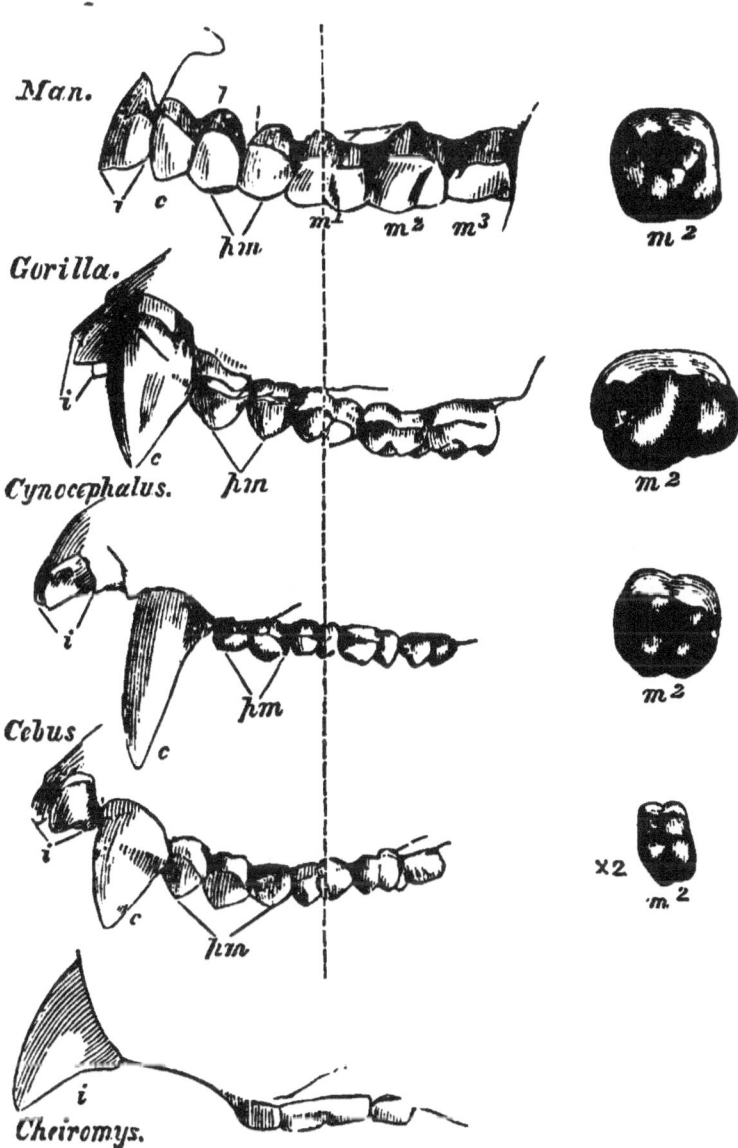

FIG. 18.—Lateral views, of the same length, of the upper jaws of various Primates. *i*, incisors; *c*, canines; *pm*, premolars; *m*, molars. A line is drawn through the first molar of Man, *Gorilla*, *Cynocephalus*, and *Cebus*, and the grinding surface of the second molar is shown in each, its anterior and internal angle being just above the *m* of *m*².

the general level of the other teeth. The roots of the false molar teeth of the Gorilla, again, are more complex than in Man, and the proportional size of the molars is different. The Gorilla has the crown of the hindmost grinder of the lower jaw more complex, and the order of eruption of the permanent teeth is different; the permanent canines making their appearance before the second and third molars in Man, and after them in the Gorilla.

Thus, while the teeth of the Gorilla closely resemble those of Man in number, kind, and in the general pattern of their crowns, they exhibit marked differences from those of Man in secondary respects, such as relative size, number of fangs, and order of appearance.

But, if the teeth of the Gorilla be compared with those of an Ape, no further removed from it than a *Cynocephalus*, or Baboon, it will be found that differences and resemblances of the same order are easily observable; but that many of the points in which the Gorilla resembles Man are those in which it differs from the Baboon; while various respects in which it differs from Man are exaggerated in the *Cynocephalus*. The number and the nature of the teeth remain the same in the Baboon as in the Gorilla and in Man. But the pattern of the Baboon's upper molars is quite different from that described above (Fig. 18), the canines are proportionally longer and more knife-like; the anterior premolar in the lower jaw is specially modified; the posterior molar of the lower jaw is still larger and more complex than in the Gorilla.

Passing from the old-world Apes to those of the new world, we meet with a change of much greater importance than any of these. In such a genus as *Cebus*, for example (Fig. 18), it will be found that while in some secondary points, such as the projection of the canines and the diastema, the resemblance to the great ape is preserved; in other and most important respects, the dentition is extremely different. Instead of 20 teeth in the milk set, there are 24: instead of 32 teeth in the permanent set, there are 36, the false molars being increased from eight to twelve. And in form the crowns of the molars are very unlike those of the Gorilla, and differ far more widely from the human pattern.

The Marmosets, on the other hand, exhibit the same number of teeth as Man and the Gorilla; but, notwithstanding this, their dentition is very different, for they have four more false molars, like the other American monkeys—but as they have four fewer true molars, the total remains the same. And passing from the American apes to the Lemurs, the dentition becomes still more completely and essentially different from that of the Gorilla. The incisors begin to vary both in number and in form. The molars acquire, more and more, a many-pointed, insectivorous character, and in one Genus, the Aye-Aye (*Cheiromys*), the canines disappear, and the teeth completely simulate those of a Rodent (Fig. 18).

Hence it is obvious that, greatly as the dentition of the highest Ape differs from that of Man, it differs far more widely from that of the lower and lowest Apes.

Whatever part of the animal fabric—whatever series of muscles, whatever viscera might be selected for comparison—the result would be the same—the lower Apes and the Gorilla would differ more than the Gorilla and the Man. I cannot attempt in this place to follow out all these comparisons in detail, and indeed it is unnecessary I should do so. But certain real, or supposed, structural distinctions between man and the apes remain, upon which so much stress has been laid, that they require careful consideration, in order that the true value may be assigned to those which are real, and the emptiness of those which are fictitious may be exposed. I refer to the characters of the hand, the foot, and the brain.

Man has been defined as the only animal possessed of two hands terminating his fore limbs, and of two feet ending his hind limbs, while it has been said that all the apes possess four hands; and he has been affirmed to differ fundamentally from all the apes in the characters of his brain, which alone, it has been strangely asserted and re-asserted, exhibits the structures known to anatomists as the posterior lobe, the posterior cornu of the lateral ventricle and the hippocampus minor.

That the former proposition should have gained general acceptance is not surprising—indeed, at first sight, appearances are much in its favour: but, as for the second, one can only admire the surpassing courage of its enunciator, seeing that it is an innovation which is not only opposed to generally and justly accepted doctrines, but which is directly negatived by the testimony of all original inquirers, who have specially investigated the matter: and that it neither has been, nor can be, supported by a single anatomical preparation. It would, in fact, be unworthy of serious refutation, except for the general and natural belief that deliberate and reiterated assertions must have some foundation.

Before we can discuss the first point with advantage we must consider with some attention, and compare together, the structure of the human hand and that of the human foot, so that we may have distinct and clear ideas of what constitutes a hand and what a foot.

The external form of the human hand is familiar enough to every one. It consists of a stout wrist followed by a broad palm, formed of flesh, and tendons, and skin,

binding together four bones, and dividing into four long
and flexible digits, or fingers, each of which bears on the
back of its last joint a broad and flattened nail. The longest cleft between any two digits is rather less than half as
long as the hand. From the outer side of the base of the
palm a stout digit goes off, having only two joints instead
of three ; so short, that it only reaches to a little beyond
the middle of the first joint of the finger next it ; and further remarkable by its great mobility, in consequence of
which it can be directed outwards, almost at a right angle
to the rest. This digit is called the '*pollex*,' or thumb ;
and, like the others, it bears a flat nail upon the back of
its terminal joint. In consequence of the proportions and
mobility of the thumb, it is what is termed " opposable ; "
in other words, its extremity can, with the greatest ease,
be brought into contact with the extremities of any of the
fingers ; a property upon which the possibility of our
carrying into effect the conceptions of the mind so largely
depends.

The external form of the foot differs widely from that
of the hand ; and yet, when closely compared, the two
present some singular resemblances. Thus the ankle corresponds in a manner with the wrist ; the sole with the
palm ; the toes with the fingers ; the great toe with the
thumb. But the toes, or digits of the foot, are far shorter
in proportion than the digits of the hand, and are less
moveable, the want of mobility being most striking in the
great toe—which, again, is very much larger in proportion to the other toes than the thumb to the fingers. In
considering this point, however, it must not be forgotten
that the civilized great toe, confined and cramped from
childhood upwards, is seen to a great disadvantage, and
that in uncivilized and barefooted people it retains a great
amount of mobility, and even some sort of opposability.

The Chinese boatmen are said to be able to pull an oar;
the artisans of Bengal to weave, and the Carajas to steal
fishhooks by its help; though, after all, it must be recollected that the structure of its joints and the arrangement
of its bones, necessarily render its prehensile action far less
perfect than that of the thumb.

But to gain a precise conception of the resemblances
and differences of the hand and foot, and of the distinctive

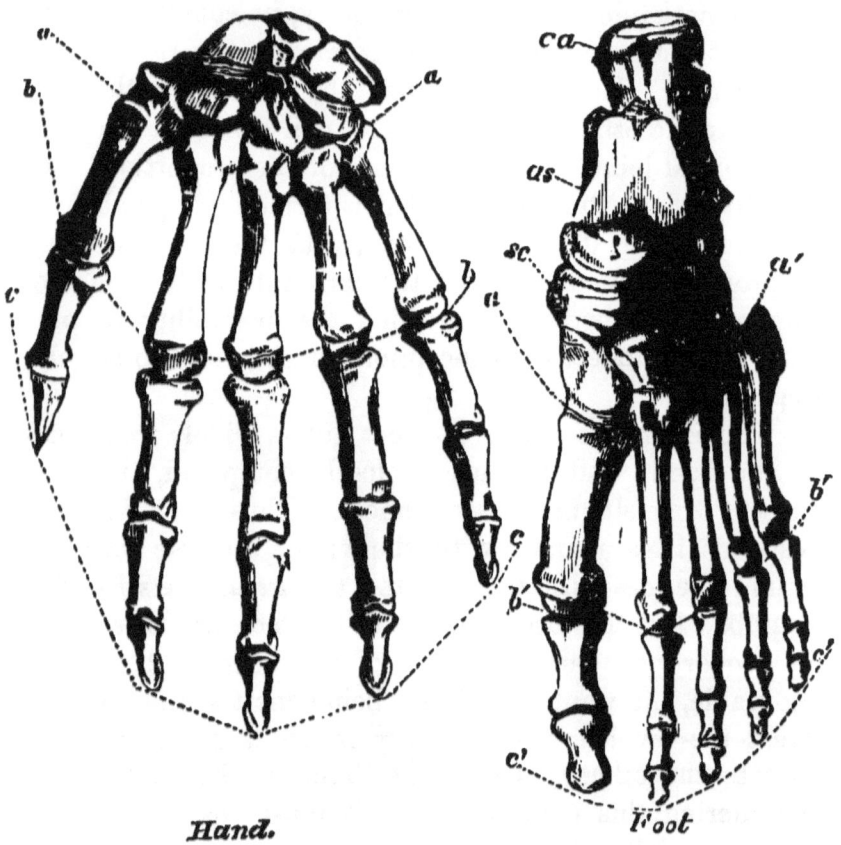

Hand.   **Foot.**

Fig. 19.—The skeleton of the Hand and Foot of Man reduced from Dr.
Carter's drawings in Gray's 'Anatomy.' The hand is drawn to a larger scale
than the foot. The line *a a* in the hand indicates the boundary between the

characters of each, we must look below the skin, and compare the bony framework and its motor apparatus in each (Fig. 19).

The skeleton of the hand exhibits, in the region which we term the wrist, and which is technically called the *carpus*—two rows of closely fitted polygonal bones, four in each row, which are tolerably equal in size. The bones of the first row with the bones of the forearm, form the wrist or joint, and are arranged side by side, no one greatly exceeding or overlapping the rest.

The four bones of the second row of the carpus bear the four long bones which support the palm of the hand. The fifth bone of the same character is articulated in a much more free and moveable manner than the others, with its carpal bone, and forms the base of the thumb. These are called *metacarpal* bones, and they carry the *phalanges*, or bones of the digits, of which there are two in the thumb and three in each of the fingers.

The skeleton of the foot is very like that of the hand in some respects. Thus there are three phalanges in each of the lesser toes, and only two in the great toe, which answers to the thumb. There is a long bone termed *metatarsal*, answering to the metacarpal, for each digit; and the *tarsus* which corresponds with the carpus, presents four short polygonal bones in a row, which correspond very closely with the four carpal bones of the second row of the hand. In other respects the foot differs very widely from the hand. Thus the great toe is the longest digit

carpus and the metacarpus; *b b* that between the latter and the proximal phalanges; *c c* marks the ends of the distal phalanges. The line *a′ a′* in the foot indicates the boundary between the tarsus and the metatarsus; *b′ b′* marks that between the metatarsus and the proximal phalanges; and *c′ c′* bounds the ends of the distal phalanges: *ca*, the calcaneum; *as*, the astragalus; *sc*, the scaphoid bone in the tarsus.

5*

but one; and its metatarsal is far less moveably articulated with the tarsus, than the metacarpal of the thumb with the carpus. But a far more important distinction lies in the fact that, instead of four more tarsal bones there are only three; and that these three are not arranged side by side, or in one row. One of them, the *os calcis* or heel bone (*ca*), lies externally, and sends back the large projecting heel; another, the *astragalus* (*as*), rests on this by one face, and by another, forms, with the bones of the leg, the ankle joint; while a third face, directed forwards, is separated from the three inner tarsal bones of the row next the metatarsus by a bone called the *scaphoid* (*sc*).

Thus there is a fundamental difference in the structure of the foot and the hand, observable when the carpus and the tarsus are contrasted; and there are differences of degree noticeable when the proportions and the mobility of the metacarpals and metatarsals, with their respective digits, are compared together.

The same two classes of differences become obvious when the muscles of the hand are compared with those of the foot.

Three principal sets of muscles, called "flexors," bend the fingers and thumb, as in clenching the fist, and three sets,—the extensors—extend them, as in straightening the fingers. These muscles are all "long muscles;" that is to say, the fleshy part of each, lying in and being fixed to the bones of the arm, is, at the other end, continued into tendons, or rounded cords, which pass into the hand, and are ultimately fixed to the bones which are to be moved. Thus, when the fingers are bent, the fleshy parts of the flexors of the fingers, placed in the arm, contract, in virtue of their peculiar endowment as muscles; and pulling the tendinous cords, connected with their ends, cause them to pull down the bones of the fingers towards the palm.

Not only are the principal flexors of the fingers and of the thumb long muscles, but they remain quite distinct from one another throughout their whole length.

In the foot, there are also three principal flexor muscles of the digits or toes, and three principal extensors; but one extensor and one flexor are short muscles; that is to say, their fleshy parts are not situated in the leg (which corresponds with the arm), but in the back and in the sole of the foot—regions which correspond with the back and the palm of the hand.

Again, the tendons of the long flexor of the toes, and of the long flexor of the great toe, when they reach the sole of the foot, do not remain distinct from one another, as the flexors in the palm of the hand do, but they become united and commingled in a very curious manner—while their united tendons receive an accessory muscle connected with the heel-bone.

But perhaps the most absolutely distinctive character about the muscles of the foot is the existence of what is termed the *peronæus longus*, a long muscle fixed to the outer bone of the leg, and sending its tendon to the outer ankle, behind and below which it passes, and then crosses the foot obliquely to be attached to the base of the great toe. No muscle in the hand exactly corresponds with this, which is eminently a foot muscle.

To resume—the foot of man is distinguished from his hand by the following absolute anatomical differences :—

1. By the arrangement of the tarsal bones.
2. By having a short flexor and a short extensor muscle of the digits.
3. By possessing the muscle termed *peronæus longus*.

And if we desire to ascertain whether the terminal
division of a limb, in other Primates, is to be called a foot
or a hand, it is by the presence or absence of these characters that we must be guided, and not by the mere proportions and greater or lesser mobility of the great toe,
which may vary indefinitely without any fundamental
alteration in the structure of the foot.

Keeping these considerations in mind, let us now turn
to the limbs of the Gorilla. The terminal division of the
fore limb presents no difficulty—bone for bone and muscle
for muscle, are found to be arranged essentially as in man,
or with such minor differences as are found as varieties in
man. The Gorilla's hand is clumsier, heavier, and has a
thumb somewhat shorter in proportion than that of man ;
but no one has ever doubted its being a true hand.

At first sight, the termination of the hind limb of the
Gorilla looks very hand-like, and as it is still more so in
many of the lower apes, it is not wonderful that the appellation " Quadrumana," or four-handed creatures, adopted
from the older anatomists* by Blumenbach, and unfortunately rendered current by Cuvier, should have gained
such wide acceptance as a name for the Simian group.
But the most cursory anatomical investigation at once

* In speaking of the foot of his "Pygmie," Tyson remarks, p. 13 :—

"But this part in the formation and in its function too, being liker a Hand
than a Foot: for the distinguishing this sort of animals from others, I have
thought whether it might not be reckoned and called rather Quadru-manus
than Quadrupes, i. e. a four-handed rather than a four-footed animal."

As this passage was published in 1699, M. I. G. St. Ililaire is clearly in
error in ascribing the invention of the term " quadrumanous" to Buffon,
though "bimanous" may belong to him. Tyson uses " Quadrumanus" in
several places, as at p. 91. . . . . " Our *Pygmie* is no Man, nor yet the
common Ape, but a sort of *Animal* between both; and though a *Biped*, yet
of the *Quadrumanus*-kind: though some *Men* too have been observed to use
their *Feet* like *Hands*, as I have seen several."

proves that the resemblance of the so-called "hind hand" to a true hand, is only skin deep, and that, in all essential respects, the hind limb of the Gorilla is as truly terminated by a foot as that of man. The tarsal bones, in all important circumstances of number, disposition, and form, resemble those of man (Fig. 20). The metatarsals and digits, on the other hand, are proportionally longer and more slender, while the great toe is not only proportionally shorter and weaker, but its metatarsal bone is united by a more moveable joint with the tarsus. At the same time, the foot is set more obliquely upon the leg than in man.

As to the muscles, there is a short flexor, a short extensor, and a *peronœus longus*, while the tendons of the long flexors of the great toe and of the other toes are united together and with an accessory fleshy bundle.

The hind limb of the Gorilla, therefore, ends in a true foot, with a very moveable great toe. It is a prehensile foot, indeed, but in no sense a hand : it is a foot which differs from that of man not in any fundamental character, but in mere proportions, in the degree of mobility, and in the secondary arrangement of its parts.

It must not be supposed, however, because I speak of these differences as not fundamental, that I wish to underrate their value. They are important enough in their way, the structure of the foot being in strict correlation with that of the rest of the organism in each case. Nor can it be doubted that the greater division of physiological labour in Man, so that the function of support is thrown wholly on the leg and foot, is an advance in the organization of very great moment to him ; but, after all, regarded anatomically, the resemblances between the foot of Man and the foot of the Gorilla are far more striking and important than the differences.

I have dwelt upon this point at length, because it is one regarding which much delusion prevails; but I might have passed it over without detriment to my argument, which only requires me to show that, be the differences between the hand and foot of Man and those of the Gorilla what they may—the differences between those of the Gorilla and those of the lower Apes are much greater.

It is not necessary to descend lower in the scale than the Orang for conclusive evidence on this head.

The thumb of the Orang differs more from that of the Gorilla than the thumb of the Gorilla differs from that of Man, not only by its shortness, but by the absence of any

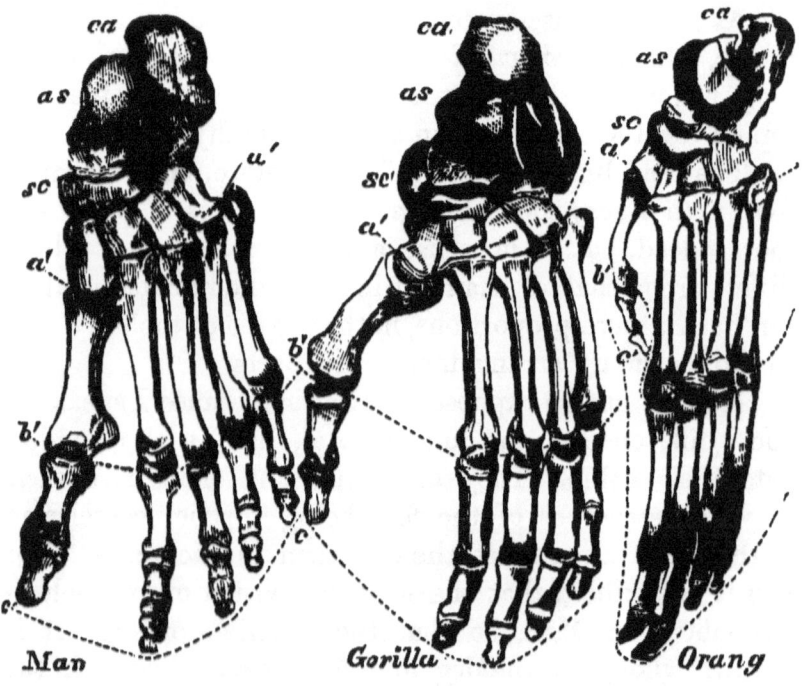

FIG. 20.—Foot of Man, Gorilla, and Orang-Utan of the same absolute length, to show the differences in proportion of each. Letters as in Fig. 19. Reduced from original drawings by Mr. Waterhouse Hawkins.

special long flexor muscle. The carpus of the Orang, like that of most lower apes, contains nine bones, while in the Gorilla, as in Man and the Chimpanzee, there are only eight.

The Orang's foot (Fig. 20) is still more aberrant; its very long toes and short tarsus, short great toe, short and raised heel, great obliquity of articulation in the leg, and absence of a long flexor tendon to the great toe, separating it far more widely from the foot of the Gorilla than the latter is separated from that of Man.

But, in some of the lower apes, the hand and foot diverge still more from those of the Gorilla, than they do in the Orang. The thumb ceases to be opposable in the American monkeys; is reduced to a mere rudiment covered by the skin in the Spider Monkey; and is directed forwards and armed with a curved claw like the other digits, in the Marmosets—so that, in all these cases, there can be no doubt but that the hand is more different from that of the Gorilla than the Gorilla's hand is from Man's.

And as to the foot, the great toe of the Marmoset is still more insignificant in proportion than that of the Orang—while in the Lemurs it is very large, and as completely thumb-like and opposable as in the Gorilla—but in these animals the second toe is often irregularly modified, and in some species the two principal bones of the tarsus, the *astragalus* and the *os calcis*, are so immensely elongated as to render the foot, so far, totally unlike that of any other mammal.

So with regard to the muscles. The short flexor of the toes of the Gorilla differs from that of Man by the circumstance that one slip of the muscle is attached, not to the heel bone, but to the tendons of the long flexors. The lower Apes depart from the Gorilla by an exaggeration of the same character, two, three, or more, slips becoming fixed to the long flexor tendons—or by a multiplication of the slips.—Again, the Gorilla differs slightly from Man in the mode of interlacing of the long flexor tendons : and the lower apes differ from the Gorilla in exhibiting yet other, sometimes very complex, arrangements of the same parts, and occasionally in the absence of the accessory fleshy bundle.

Throughout all these modifications it must be recollected that the foot loses no one of its essential characters. Every Monkey and Lemur exhibits the characteristic arrangement of tarsal bones, possesses a short flexor and short extensor muscle, and a *peronæus longus*. Varied as the proportions and appearance of the organ may be, the terminal division of the hind limb remains, in plan and principle of construction, a foot, and never, in those respects, can be confounded with a hand.

Hardly any part of the bodily frame, then, could be found better calculated to illustrate the truth that the structural differences between Man and the highest Ape are of less value than those between the highest and the lower Apes, than the hand or the foot, and yet, perhaps, there is one organ the study of which enforces the same conclusion in a still more striking manner—and that is the Brain.

But before entering upon the precise question of the amount of difference between the Ape's brain and that of Man, it is necessary that we should clearly understand what constitutes a great, and what a small difference in cerebral structure ; and we shall be best enabled to do this by a brief study of the chief modifications which the brain exhibits in the series of vertebrate animals.

The brain of a fish is very small, compared with the spinal cord into which it is continued, and with the nerves which come off from it : of the segments of which it is composed—the olfactory lobes, the cerebral hemisphere, and the succeeding divisions—no one predominates so much over the rest as to obscure or cover them ; and the so-called optic lobes are, frequently, the largest masses of all. In Reptiles, the mass of the brain, relatively to the spinal cord, increases and the cerebral hemispheres begin to predominate over the other parts ; while in Birds this predominance is still more marked. The brain of the lowest Mammals, such as the duck-billed Platypus and the Opossums and Kangaroos, exhibits a still more definite advance in the same direction. The cerebral hemispheres have now so much increased in size as, more or less, to hide the representatives of the optic lobes, which remain comparatively small, so that the brain of a Marsupial is extremely different from that of a Bird, Reptile, or Fish. A step higher in the scale, among the placental Mammals, the structure of the brain acquires a vast modification— not that it appears much altered externally, in a Rat or in a Rabbit, from what it is in a Marsupial—nor that the proportions of its parts are much changed, but an apparently new structure is found between the cerebral hemispheres, connecting them together, as what is called the 'great commissure' or 'corpus callosum.' The subject requires careful re-investigation, but if the currently received statements are correct, the appearance of the 'corpus callosum' in the placental mammals is the greatest and most sudden modification exhibited by the brain in the whole series of vertebrated animals—it is the greatest leap anywhere made by Nature in her brain work. For the two halves of the brain being once thus knit together, the progress of cerebral complexity is traceable through a complete series of steps from the lowest Rodent, or Insectivore, to Man ; and that complexity consists, chiefly, in the disproportionate development of the cerebral hemispheres and of the cerebellum, but especially of the former, in respect to the other parts of the brain.

In the lower placental mammals, the cerebral hemispheres leave the proper upper and posterior face of the cerebellum completely visible, when the brain is viewed from above, but, in the higher forms, the hinder part of each hemisphere, separated only by the tentorium (p. 117) from the anterior face of the cerebellum, inclines backwards and downwards, and grows out, as the so-called "posterior lobe," so as at length to overlap and hide the cerebellum. In all Mammals, each cerebral hemisphere contains a cavity which is termed the 'ventricle' and as this ventricle is prolonged, on the one hand, forwards, and on the other downwards, into the substance of the hemisphere, it is said to have two horns or 'cornua,' an 'anterior cornu,' and a 'descending cornu.' When the posterior lobe is well developed, a third prolongation of the ventricular cavity extends into it, and is called the "posterior cornu."

In the lower and smaller forms of placental Mammals the surface of the cerebral hemispheres is either smooth or evenly rounded, or exhibits a very few grooves, which are technically termed 'sulci,' separating ridges or 'convolutions' of the substance of the brain ; and the smaller species of all orders tend to a similar smoothness of brain. But in the higher orders, and especially the larger members of these orders, the grooves, or sulci, become extremely numerous, and the intermediate convolutions proportionately more complicated in their meanderings, until, in the Elephant, the Porpoise, the higher Apes, and Man, the cerebral surface appears a perfect labyrinth of tortuous foldings.

Where a posterior lobe exists and presents its customary cavity—the posterior cornu—it commonly happens that a particular sulcus appears upon the inner and under surface of the lobe, parallel with and beneath the floor of the cornu—which is, as it were, arched over the roof of the sulcus. It is as if the groove had been formed by indenting the floor of the posterior horn from without with a blunt instrument, so that the floor should rise as a convex eminence. Now this eminence is what has been termed the 'Hippocampus minor;' the 'Hippocampus major' being a larger eminence in the floor of the descending cornu. What may be the functional importance of either of these structures we know not.

As if to demonstrate, by a striking example, the impossibility of erecting any cerebral barrier between man and the apes, Nature has provided us, in the latter animals, with an almost complete series of gradations from brains little higher than that of a Rodent, to brains little lower than that of Man. And it is a remarkable circumstance; that though, so far as our present knowledge extends, there *is* one true structural break in the series of forms of Simian brains, this hiatus does not lie between Man and the man-like apes, but between the lower and the lowest Simians; or, in other words, between the old and new world apes and monkeys, and the Lemurs. Every Lemur which has yet been examined, in fact, has its cerebellum partially visible from above, and its posterior lobe, with the contained posterior cornu and hippocampus minor, more or less rudimentary. Every Marmoset, American monkey, old world monkey, Baboon, or Man-like ape, on the contrary, has its cerebellum entirely hidden, posteriorly, by the cerebral lobes, and possesses a large posterior cornu, with a well developed hippocampus minor.

In many of these creatures, such as the Saimiri (*Chry-*

sothrix), the cerebral lobes overlap and extend much further behind the cerebellum, in proportion, than they do in man (Fig. 17)—and it is quite certain that, in all, the cerebellum is completely covered behind, by well developed posterior lobes. The fact can be verified by every one who possesses the skull of any old or new world monkey. For, inasmuch as the brain in all mammals completely fills the cranial cavity, it is obvious that a cast of the interior of the skull will reproduce the general form of the brain, at any rate with such minute and, for the present purpose, utterly unimportant differences as may result from the absence of the enveloping membranes of the brain in the dry skull. But if such a cast be made in plaster, and compared with a similar cast of the interior of a human skull, it will be obvious that the cast of the cerebral chamber, representing the cerebrum of the ape, as completely covers over and overlaps the cast of the cerebellar chamber, representing the cerebellum, as it does in the man (Fig. 21). A careless observer, forgetting that a soft structure like the brain loses its proper shape the moment it is taken out of the skull, may indeed mistake the uncovered condition of the cerebellum of an extracted and distorted brain for the natural relations of the parts; but his error must become patent even to himself if he try to replace the brain within the cranial chamber. To suppose that the cerebellum of an ape is naturally uncovered behind is a miscomprehension comparable only to that of one who should imagine that a man's lungs always occupy but a small portion of the thoracic cavity—because they do so when the chest is opened, and their elasticity is no longer neutralized by the pressure of the air.

And the error is the less excusable, as it must become apparent to every one who examines a section of the skull of any ape above a Lemur, without taking the trouble to make a cast of it. For there is a very marked groove in every such skull, as in the human skull—which indicates the line of attachment of what is termed the *tentorium*—a sort of parchment-like shelf, or partition, which,

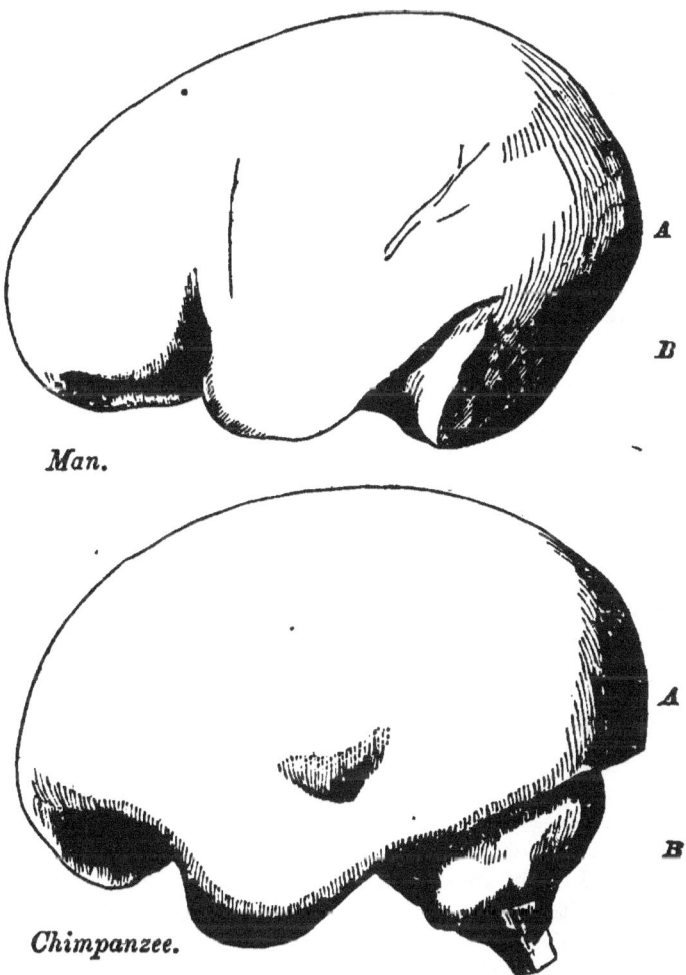

FIG. 21.—Drawings of the internal casts of a Man's and of a Chimpanzee's skull, of the same absolute length, and placed in corresponding positions, *A*. Cerebrum ; *B*. Cerebellum. The former drawing is taken from a cast in the Museum of the Royal College of Surgeons, the latter from the photograph

in the recent state, is interposed between the cerebrum and the cerebellum, and prevents the former from pressing upon the latter (see Fig. 17).

This groove, therefore, indicates the line of separation between that part of the cranial cavity which contains the cerebrum, and that which contains the cerebellum ; and as the brain exactly fills the cavity of the skull, it is obvious that the relations of these two parts of the cranial cavity at once informs us of the relations of their contents. Now in man, in all the old world, and in all the new world Simiæ, with one exception, when the face is directed forwards, this line of attachment of the tentorium, or impression for the lateral sinus, as it is technically called, is nearly horizontal, and the cerebral chamber invariably overlaps or projects behind the cerebellar chamber. In the Howler Monkey or *Mycetes* (see Fig. 17), the line passes obliquely upwards and backwards, and the cerebral overlap is almost nil ; while in the Lemurs, as in the lower mammals, the line is much more inclined in the same direction, and the cerebellar chamber projects considerably beyond the cerebral.

When the gravest errors respecting points so easily settled as this question respecting the posterior lobes, can be authoritatively propounded, it is no wonder that matters of observation, of no very complex character, but still requiring a certain amount of care, should have fared worse. Any one who cannot see the posterior lobe in an

of the cast of a Chimpanzee's skull, which illustrates the paper by Mr. Marshall 'On the Brain of the Chimpanzee' in the Natural History Review for July, 1861. The sharper definition of the lower edge of the cast of the cerebral chamber in the Chimpanzee arises from the circumstance that the tentorium remained in that skull and not in the Man's. The cast more accurately represents the brain in the Chimpanzee than in Man ; and the great backward projection of the posterior lobes of the cerebrum of the former, beyond the cerebellum, is conspicuous.

ape's brain is not likely to give a very valuable opinion respecting the posterior cornu or the hippocampus minor. If a man cannot see a church, it is preposterous to take his opinion about its altar-piece or painted window—so that I do not feel bound to enter upon any discussion of these points, but content myself with assuring the reader that the posterior cornu and the hippocampus minor, have now been seen—usually, at least as well developed as in man, and often better—not only in the Chimpanzee, the Orang, and the Gibbon, but in all the genera of the old world baboons and monkeys, and in most of the new world forms, including the Marmosets.*

In fact, all the abundant and trustworthy evidence (consisting of the results of careful investigations directed to the determination of these very questions, by skilled anatomists) which we now possess, leads to the conviction that, so far from the posterior lobe, the posterior cornu, and the hippocampus minor, being structures peculiar to and characteristic of man, as they have been over and over again asserted to be, even after the publication of the clearest demonstration of the reverse, it is precisely these structures which are the most marked cerebral characters common to man with the apes. They are among the most distinctly Simian peculiarities which the human organism exhibits.

As to the convolutions, the brains of the apes exhibit every stage of progress, from the almost smooth brain of the Marmoset, to the Orang and the Chimpanzee, which fall but little below Man. And it is most remarkable that, as soon as all the principal sulci appear, the pattern according to which they are arranged is identical with that of the corresponding sulci of man. The surface of

* See the note at the end of this essay for a succinct history of the controversy to which allusion is here made.

the brain of a monkey exhibits a sort of skeleton map of
man's, and in the man-like apes the details become more
and more filled in, until it is only in minor characters,
such as the greater excavation of the anterior lobes, the
constant presence of fissures usually absent in man, and
the different disposition and proportions of some convolutions, that the Chimpanzee's or the Orang's brain can be
structurally distinguished from Man's.

So far as cerebral structure goes, therefore, it is clear
that Man differs less from the Chimpanzee or the Orang,
than these do even from the Monkeys, and that the difference between the brains of the Chimpanzee and of Man is
almost insignificant, when compared with that between the
Chimpanzee brain and that of a Lemur.

It must not be overlooked, however, that there is a
very striking difference in absolute mass and weight between the lowest human brain and that of the highest ape
—a difference which is all the more remarkable when we
recollect that a full grown Gorilla is probably pretty nearly
twice as heavy as a Bosjes man, or as many an European
woman. It may be doubted whether a healthy human
adult brain ever weighed less than thirty-one or -two
ounces, or that the heaviest Gorilla brain has exceeded
twenty ounces.

This is a very noteworthy circumstance, and doubtless
will one day help to furnish an explanation of the great
gulf which intervenes between the lowest man and the
highest ape in intellectual power ;* but it has little sys-

---

* I say *help* to furnish: for I by no means believe that it was any original
difference of cerebral quality, or quantity, which caused that divergence between the human and the pithecoid stirpes, which has ended in the present
enormous gulf between them. It is no doubt perfectly true, in a certain sense,
that all difference of function is a result of difference of structure; or, in other
words, of difference in the combination of the primary molecular forces of
living substance; and, starting from this undeniable axiom, objectors occasion-

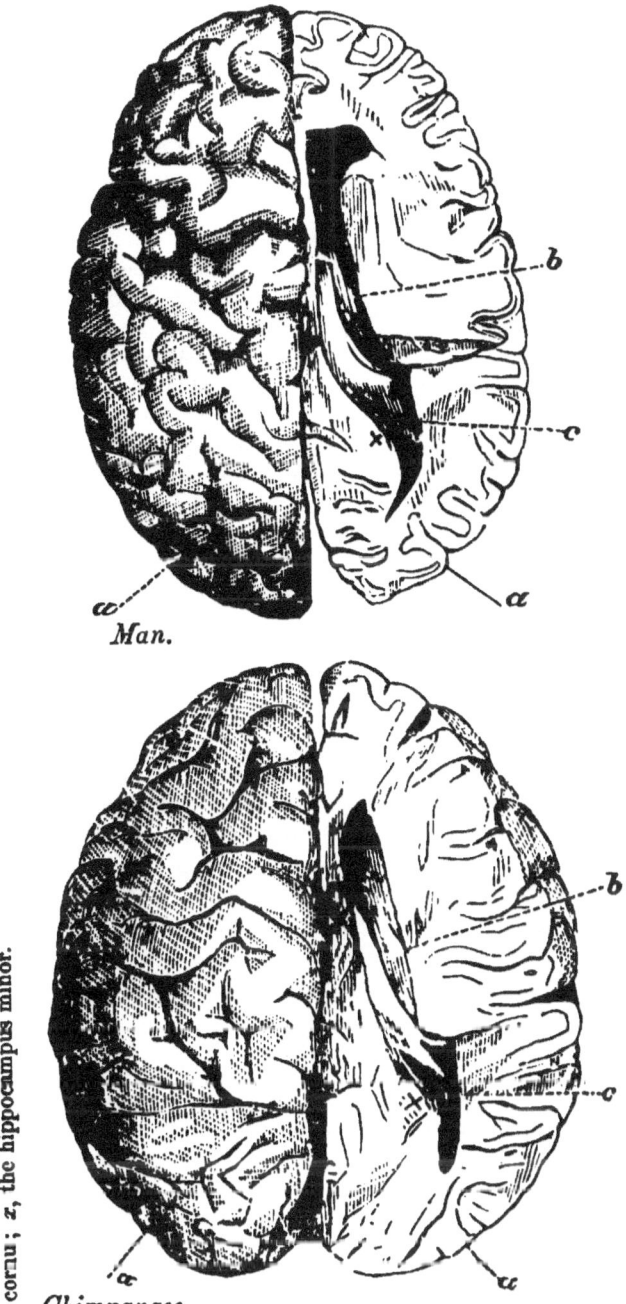

FIG. 22.—Drawings of the cerebral hemispheres of a Man and of a Chimpanzee of the same length, in order to show the relative proportions of the parts; the former taken from a specimen, which Mr. Flower, Conservator of the Museum of the Royal College of Surgeons, was good enough to dissect for me; the latter, from the photograph of a similarly dissected Chimpanzee's brain, given in Mr. Marshall's paper above referred to. *a*, posterior lobe; *b*, lateral ventricle; *c*, posterior cornu; *z*, the hippocampus minor.

Man.

Chimpanzee.

tematic value, for the simple reason that, as may be concluded from what has been already said respecting cranial capacity, the difference in weight of brain between the highest
and the lowest men is far greater, both relatively and absolutely, than that between the lowest man and the highest
ape.   The latter, as has been seen, is represented by, say

ally, and with much seeming plausibility, argue that the vast intellectual chasm
between the Ape and Man implies a corresponding structural chasm in the
organs of the intellectual functions ; so that, it is said, the non-discovery of
such vast differences proves, not that they are absent, but that Science is incompetent to detect them.   A very little consideration, however, will, I think,
show the fallacy of this reasoning.   Its validity hangs upon the assumption,
that intellectual power depends altogether on the brain—whereas the brain is
only one condition out of many on which intellectual manifestations depend ;
the others being, chiefly, the organs of the senses and the motor apparatuses,
especially those which are concerned in prehension and in the production of
articulate speech.

A man born dumb, notwithstanding his great cerebral mass and his inherit-
ance of strong intellectual instincts, would be capable of few higher intellectual manifestations than an Orang or a Chimpanzee, if he were confined to the
society of dumb associates.   And yet there might not be the slightest discernible difference between his brain and that of a highly intelligent and cultivated
person.   The dumbness might be the result of a defective structure of the
mouth, or of the tongue, or a mere defective innervation of these parts ; or it
might result from congenital deafness, caused by some minute defect of the
internal ear, which only a careful anatomist could discover.

The argument, that because there is an immense difference between a Man's
intelligence and an Ape's, therefore, there must be an equally immense difference between their brains, appears to me to be about as well based as the
reasoning by which one should endeavour to prove that, because there is a
" great gulf" between a watch that keeps accurate time and another that will
not go at all, there is therefore a great structural hiatus between the two
watches.   A hair in the balance-wheel, a little rust on a pinion, a bend in a
tooth of the escapement, a something so slight that only the practised eye of
the watchmaker can discover it, may be the source of all the difference.

And believing, as I do, with Cuvier, that the possession of articulate speech
is the grand distinctive character of man (whether it be absolutely peculiar to
him or not), I find it very easy to comprehend, that some equally inconspi-
cuous structural difference may have been the primary cause of the immeasu-
rable and practically infinite divergence of the Human from the Simian Stirps.

twelve, ounces of cerebral substance absolutely, or by 32 :
20 relatively; but as the largest recorded human brain
weighed between 65 and 66 ounces, the former difference
is represented by more than 33 ounces absolutely, or by
65 : 32 relatively. Regarded systematically, the cerebral
differences, of man and apes, are not of more than generic
value—his Family distinction resting chiefly on his dentition, his pelvis, and his lower limbs.

Thus, whatever system of organs be studied, the comparison of their modifications in the ape series leads to one and
the same result—that the structural differences which separate Man from the Gorilla and the Chimpanzee are not
so great as those which separate the Gorilla from the
lower apes.

But in enunciating this important truth I must guard
myself against a form of misunderstanding, which is very
prevalent. I find, in fact, that those who endeavour to
teach what nature so clearly shows us in this matter, are
liable to have their opinions misrepresented and their
phraseology garbled, until they seem to say that the structural differences between man and even the highest apes
are small and insignificant. Let me take this opportunity
then of distinctly asserting, on the contrary, that they are
great and significant; that every bone of a Gorilla bears
marks by which it might be distinguished from the corresponding bone of a Man; and that, in the present creation, at any rate, no intermediate link bridges over the
gap between *Homo* and *Troglodytes*.

It would be no less wrong than absurd to deny the existence of this chasm; but it is at least equally wrong and
absurd to exaggerate its magnitude, and, resting on the
admitted fact of its existence, to refuse to inquire whether
it is wide or narrow. Remember, if you will, that there

is no existing link between Man and the Gorilla, but do
not forget that there is a no less sharp line of demarcation, a no less complete absence of any transitional form,
between the Gorilla and the Orang, or the Orang and the
Gibbon.   I say, not less sharp, though it is somewhat nar-
rower.   The structural differences between Man and the
Man-like apes certainly justify our regarding him as con-
stituting a family apart from them ; though, inasmuch as
he differs less from them than they do from other families
of the same order, there can be no justification for placing
him in a distinct order.

And thus the sagacious foresight of the great lawgiver
of systematic zoology, Linnæus, becomes justified, and a
century of anatomical research brings us back to his conclusion, that man is a member of the same order (for which
the Linnæan term PRIMATES ought to be retained) as the
Apes and Lemurs.   This order is now divisible into seven
families, of about equal systematic value : the first, the
ANTHROPINI, contains Man alone ; the second, the CA-
TARHINI, embraces the old world apes ; the third, the
PLATYRHINI, all new world apes, except the Marmosets ;
the fourth, the ARCTOPITHECINI, contains the Marmosets ;
the fifth, the LEMURINI, the Lemurs—from which *Cheiromys* should probably be excluded to form a sixth distinct family, the CHEIROMYINI ; while the seventh, the
GALEOPITHECINI, contains only the flying Lemur *Galeo-
pithecus,*—a strange form which almost touches on the
Bats, as the *Cheiromys* puts on a Rodent clothing, and
the Lemurs simulate Insectivora.

Perhaps no order of mammals presents us with so extraordinary a series of gradations as this—leading us in-
sensibly from the crown and summit of the animal creation down to creatures, from which there is but a step, as
it seems, to the lowest, smallest, and least intelligent of

the placental Mammalia. It is as if nature herself had foreseen the arrogance of man, and with Roman severity had provided that his intellect, by its very triumphs, should call into prominence the slaves, admonishing the conqueror that he is but dust.

These are the chief facts, this the immediate conclusion from them to which I adverted in the commencement of this Essay. The facts, I believe, cannot be disputed; and if so, the conclusion appears to me to be inevitable.

But if Man be separated by no greater structural barrier from the brutes than they are from one another—then it seems to follow that if any process of physical causation can be discovered by which the genera and families of ordinary animals have been produced, that process of causation is amply sufficient to account for the origin of Man. In other words, if it could be shown that the Marmosets, for example, have arisen by gradual modification of the ordinary Platyrhini, or that both Marmosets and Platyrhini are modified ramifications of a primitive stock —then, there would be no rational ground for doubting that man might have originated, in the one case, by the gradual modification of a man-like ape ; or, in the other case, as a ramification of the same primitive stock as those apes.

At the present moment, but one such process of physical causation has any evidence in its favour ; or, in other words, there is but one hypothesis regarding the origin of species of animals in general which has any scientific existence—that propounded by Mr. Darwin. For Lamarck, sagacious as many of his views were, mingled them with so much that was crude and even absurd, as to neutralize the benefit which his originality might have effected, had he been a more sober and cautious thinker ; and though

I have heard of the announcement of a formula touching "the ordained continuous becoming of organic forms," it is obvious that it is the first duty of a hypothesis to be intelligible, and that a qua-quâ-versal proposition of this kind, which may be read backwards, or forwards, or sideways, with exactly the same amount of signification, does not really exist, though it may seem to do so.

At the present moment, therefore, the question of the relation of man to the lower animals resolves itself, in the end, into the larger question of the tenability or untenability of Mr. Darwin's views. But here we enter upon difficult ground, and it behoves us to define our exact position with the greatest care.

It cannot be doubted, I think, that Mr. Darwin has satisfactorily proved that what he terms selection, or selective modification, must occur, and does occur, in nature; and he has also proved to superfluity that such selection is competent to produce forms as distinct, structurally, as some genera even are. If the animated world presented us with none but structural differences, I should have no hesitation in saying that Mr. Darwin has demonstrated the existence of a true physical cause, amply competent to account for the origin of living species, and of man among the rest.

But, in addition to their structural distinctions, the species of animals and plants, or at least a great number of them, exhibit physiological characters—what are known as distinct species, structurally, being for the most part either altogether incompetent to breed one with another; or if they breed, the resulting mule, or hybrid, is unable to perpetuate its race with another hybrid of the same kind.

A true physical cause is, however, admitted to be such only on one condition—that it shall account for all the phenomena which come within the range of its operation.

If it is inconsistent with any one phenomenon, it must be rejected; if it fails to explain any one phenomenon, it is so far weak, so far to be suspected; though it may have a perfect right to claim provisional acceptance.

Now, Mr. Darwin's hypothesis is not, so far as I am aware, inconsistent with any known biological fact; on the contrary, if admitted, the facts of Development, of Comparative Anatomy, of Geographical Distribution, and of Palæontology, become connected together, and exhibit a meaning such as they never possessed before; and I, for one, am fully convinced that if not precisely true, that hypothesis is as near an approximation to the truth as, for example, the Copernican hypothesis was to the true theory of the planetary motions.

But, for all this, our acceptance of the Darwinian hypothesis must be provisional so long as one link in the chain of evidence is wanting; and so long as all the animals and plants certainly produced by selective breeding from a common stock are fertile, and their progeny are fertile with one another, that link will be wanting. For, so long, selective breeding will not be proved to be competent to do all that is required of it to produce natural species.

I have put this conclusion as strongly as possible before the reader, because the last position in which I wish to find myself is that of an advocate for Mr. Darwin's, or any other views—if by an advocate is meant one whose business it is to smooth over real difficulties, and to persuade where he cannot convince.

In justice to Mr. Darwin, however, it must be admitted that the conditions of fertility and sterility are very ill understood, and that every day's advance in knowledge leads us to regard the hiatus in his evidence as of less and less importance, when set against the multitude of facts which harmonize with, or receive an explanation from, his doctrines.

I adopt Mr. Darwin's hypothesis, therefore, subject to the production of proof that physiological species may be produced by selective breeding; just as a physical philosopher may accept the undulatory theory of light, subject to the proof of the existence of the hypothetical ether; or as the chemist adopts the atomic theory, subject to the proof of the existence of atoms; and for exactly the same reasons, namely, that it has an immense amount of primâ facie probability: that it is the only means at present within reach of reducing the chaos of observed facts to order; and lastly, that it is the most powerful instrument of investigation which has been presented to naturalists since the invention of the natural system of classification and the commencement of the systematic study of embryology.

But even leaving Mr. Darwin's views aside, the whole analogy of natural operations furnishes so complete and crushing an argument against the intervention of any but what are termed secondary causes, in the production of all the phenomena of the universe; that, in view of the intimate relations between Man and the rest of the living world; and between the forces exerted by the latter and all other forces, I can see no excuse for doubting that all are co-ordinated terms of Nature's great progression, from the formless to the formed—from the inorganic to the organic—from blind force to conscious intellect and will.

Science has fulfilled her function when she has ascertained and enunciated truth; and were these pages addressed to men of science only, I should now close this essay, knowing that my colleagues have learned to respect nothing but evidence, and to believe that their highest duty lies in submitting to it, however it may jar against their inclinations.

But desiring, as I do, to reach the wider circle of the intelligent public, it would be unworthy cowardice were I to ignore the repugnance with which the majority of my readers are likely to meet the conclusions to which the most careful and conscientious study I have been able to give to this matter, has led me.

On all sides I shall hear the cry—" We are men and women, and not a mere better sort of apes, a little longer in the leg, more compact in the foot, and bigger in brain than your brutal Chimpanzees and Gorillas. The power of knowledge—the conscience of good and evil—the pitiful tenderness of human affections, raise us out of all real fellowship with the brutes, however closely they may seem to approximate us."

To this I can only reply that the exclamation would be most just and would have my own entire sympathy, if it were only relevant. But it is not I who seek to base Man's dignity upon his great toe, or insinuate that we are lost if an Ape has a hippocampus minor. On the contrary, I have done my best to sweep away this vanity. I have endeavoured to show that no absolute structural line of demarcation, wider than that between the animals which immediately succeed us in the scale, can be drawn between the animal world and ourselves; and I may add the expression of my belief that the attempt to draw a physical distinction is equally futile, and that even the highest faculties of feeling and of intellect begin to germinate in lower forms of life.* At the same time no one is

* It is so rare a pleasure for me to find Professor Owen's opinions in entire accordance with my own, that I cannot forbear from quoting a paragraph which appeared in his Essay " On the Characters, &c. of the Class Mammalia," in the 'Journal of the Proceedings of the Linnean Society of London' for

6*

more strongly convinced than I am of the vastness of the gulf between civilized man and the brutes; or is more certain that whether *from* them or not, he is assuredly not *of* them. No one is less disposed to think lightly of the present dignity, or despairingly of the future hopes, of the only consciously intelligent denizen of this world.

We are indeed told by those who assume authority in these matters, that the two sets of opinions are incompatible, and that the belief in the unity of origin of man and brutes involves the brutalization and degradation of the former. But is this really so? Could not a sensible child confute, by obvious arguments, the shallow rhetoricians who would force this conclusion upon us? Is it, indeed, true, that the Poet, or the Philosopher, or the Artist whose genius is the glory of his age, is degraded from his high estate by the undoubted historical probability, not to say certainty, that he is the direct descendant of some naked and bestial savage, whose intelligence was just sufficient to make him a little more cunning than the Fox, and by so much more dangerous than the Tiger? Or is he bound to howl and grovel on all fours because of the wholly unquestionable fact, that he was once an egg, which

1857, but is unaccountably omitted in the "Reade Lecture" delivered before the University of Cambridge two years later, which is otherwise nearly a reprint of the paper in question. Prof. Owen writes:

"Not being able to appreciate or conceive of the distinction between the psychical phenomena of a Chimpanzee and of a Boschisman or of an Aztec, with arrested brain growth, as being of a nature so essential as to preclude a comparison between them, or as being other than a difference of degree, I cannot shut my eyes to the significance of that all-pervading similitude of structure—every tooth, every bone, strictly homologous—which makes the determination of the difference between *Homo* and *Pithecus* the anatomist's difficulty."

Surely it is a little singular, that the 'anatomist,' who finds it 'difficult' to 'determine the difference' between *Homo* and *Pithecus*, should yet range them on anatomical grounds, in distinct sub-classes!

no ordinary power of discrimination could distinguish from
that of a Dog? Or is the philanthropist or the saint to
give up his endeavours to lead a noble life, because the
simplest study of man's nature reveals, at its foundations,
all the selfish passions and fierce appetites of the merest
quadruped? Is mother-love vile because a hen shows it,
or fidelity base because dogs possess it?

The common sense of the mass of mankind will answer
these questions without a moment's hesitation. Healthy
humanity, finding itself hard pressed to escape from real
sin and degradation, will leave the brooding over specula-
tive pollution to the cynics and the righteous 'overmuch'
who, disagreeing in everything else, unite in blind insen-
sibility to the nobleness of the visible world, and in ina-
bility to appreciate the grandeur of the place Man occupies therein.

Nay more, thoughtful men, once escaped from the
blinding influences of traditional prejudice, will find in the
lowly stock whence man has sprung, the best evidence of
the splendour of his capacities; and will discern in his
long progress through the Past, a reasonable ground of
faith in his attainment of a nobler Future.

They will remember that in comparing civilized man
with the animal world, one is as the Alpine traveller, who
sees the mountains soaring into the sky and can hardly
discern where the deep shadowed crags and roseate peaks
end, and where the clouds of heaven begin. Surely the
awe-struck voyager may be excused if, at first, he refuses
to believe the geologist, who tells him that these glorious
masses are, after all, the hardened mud of primeval seas,
or the cooled slag of subterranean furnaces—of one substance with the dullest clay, but raised by inward forces
to that place of proud and seemingly inaccessible glory.

But the geologist is right; and due reflection on his

teachings, instead of diminishing our reverence and our wonder, adds all the force of intellectual sublimity, to the mere æsthetic intuition of the uninstructed beholder.

And after passion and prejudice have died away, the same result will attend the teachings of the naturalist respecting that great Alps and Andes of the living world—Man. Our reverence for the nobility of manhood will not be lessened by the knowledge, that Man is, in substance and in structure, one with the brutes ; for, he alone possesses the marvellous endowment of intelligible and rational speech, whereby, in the secular period of his existence, he has slowly accumulated and organized the experience which is almost wholly lost with the cessation of every individual life in other animals ; so that now he stands raised upon it as on a mountain top, far above the level of his humble fellows, and transfigured from his grosser nature by reflecting, here and there, a ray from the infinite source of truth.

A SUCCINCT HISTORY OF THE CONTROVERSY RESPECTING THE CEREBRAL STRUCTURE OF MAN AND THE APES.

Up to the year 1857 all anatomists of authority, who had occupied themselves with the cerebral structure of the Apes—Cuvier, Tiedemann, Sandifort, Vrolik, Isidore G. St. Hilaire, Schroeder van der Kolk, Gratiolet—were agreed that the brain of the Ape possesses a POSTERIOR LOBE.

Tiedemann, in 1825, figured and acknowledged in the text of his 'Icones,' the existence of the POSTERIOR CORNU of the lateral ventricle in the Apes, not only under the title of ' Scrobiculus parvus loco cornu posterioris'—a fact which has been paraded—but as ' cornu posterius' (Icones, p. 54), a circumstance which has been, as sedulously, kept in the back ground.

Cuvier (Lecons, T. iii. p. 103) says, "the anterior or lateral ventricles possess a digital cavity [posterior cornu] only in Man and the Apes . . . . . Its presence depends on that of the posterior lobes."

Schroeder van der Kolk and Vrolik, and Gratiolet, had also figured and described the posterior cornu in various Apes. As to the HIPPOCAMPUS MINOR, Tiedemann had erroneously asserted its absence in the Apes; but Schroeder van der Kolk and Vrolik had pointed out the existence of what they considered a rudimentary one in the Chimpanzee, and Gratiolet had expressly affirmed its existence in these animals. Such was the state of our information on these subjects in the year 1856.

In the year 1857, however, Professor Owen, either in ignorance of these well-known facts or else unjustifiably suppressing them, submitted to the Linnæan Society a paper " On the Characters, Principles of Division, and Primary Groups of the Class Mammalia," which was printed in the Society's Journal, and contains the following passage :—" In Man, the brain presents an ascensive step in development, higher and more strongly marked than that by which the preceding sub-class was distinguished from the one below it.   Not only do the cerebral hemispheres overlap the olfactory lobes and cerebellum, but they extend in advance of the one and further back than the other.   The posterior development is so marked, that anatomists have assigned to that part the character of a third lobe ; *it is peculiar to the genus Homo, and equally peculiar is the posterior horn of the lateral ventricle and the 'hippocampus minor' which characterise the hind lobe of each hemisphere.*"—*Journal of the Proceedings of the Linnæan Society,* Vol. ii. p. 19.

As the essay in which this passage stands had no less ambitious an aim than the remodelling of the classification of the Mammalia, its author might be supposed to have written under a sense of peculiar responsibility, and to have tested, with especial care, the statements he ventured to promulgate.   And even if this be expecting too much, hastiness, or want of opportunity for due deliberation, cannot now be pleaded in extenuation of any shortcomings ; for the propositions cited were repeated two years afterwards in the Reade Lecture, delivered before so grave a body as the University of Cambridge, in 1859.

When the assertions, which I have italicised in the above extract, first came under my notice, I was not a little astonished at so flat a contradiction of the doctrines current among well-informed anatomists ; but, not unnaturally imagining that the deliberate statements of a responsible person must have some foundation in fact, I deemed it my duty to investigate the subject anew before the time at which it would be my business to lecture thereupon came round. The result of my inquiries was to prove that Mr. Owen's three assertions, that " the third lobe, the posterior horn of the lateral ventricle, and the hippocampus minor," are " peculiar to the genus *Homo*," are contrary to the plainest facts.   I communicated this conclusion to the students of my class ; and then, having no desire to embark in a controversy which could not redound to the honour of British science, whatever its issue, I turned to more congenial occupations.

The time speedily arrived, however, when a persistence in this reticence would have involved me in an unworthy paltering with truth.

At the meeting of the British Association at Oxford, in 1860, Professor Owen repeated these assertions in my presence, and, of course, I immediately gave them a direct and unqualified contradiction, pledging myself to justify that unusual procedure elsewhere.

I redeemed that pledge by publishing, in the January number of the *Natural History Review* for 1861, an article wherein the truth of the three following propositions was fully demonstrated (*l. c.* p. 71) :—

" 1. That the third lobe is neither peculiar to, nor characteristic of, man, seeing that it exists in all the higher quadrumana."

" 2. That the posterior cornu of the lateral ventricle is neither peculiar to, nor characteristic of, man, inasmuch as it also exists in the higher quadrumana.

" 3. That the *hippocampus minor* is neither peculiar to, nor characteristic of, man, as it is found in certain of the higher quadrumana."

Furthermore, this paper contains the following paragraph (p. 76) :

" And lastly, Schroeder van der Kolk and Vrolik (op. cit. p. 271), though they particularly note that 'the lateral ventricle is distinguished from that of Man by the very defective proportions of the posterior cornu, wherein only a stripe is visible as an indication of the hippocampus minor ;' yet the Figure 4, in their second Plate, shows that this posterior cornu is a perfectly distinct and unmistakeable structure, quite as large as it often is in Man. It is the more remarkable that Professor Owen should have overlooked the explicit statement and figure of these authors, as it is quite obvious, on comparison of the figures, that his woodcut of the brain of a Chimpanzee (l. c. p. 19) is a reduced copy of the second figure of Messrs. Schroeder van der Kolk and Vrolik's first Plate.

" As M. Gratiolet (l. c. p. 18), however, is careful to remark, ' unfortunately the brain which they have taken as a model was greatly altered (profondément affaissé), whence the general form of the brain is given in these plates in a manner which is altogether incorrect.' Indeed, it is perfectly obvious, from a comparison of a section of the skull of the Chimpanzee with these figures, that such is the case ; and it is greatly to be regretted that so inadequate a figure should have been taken as a typical representation of the Chimpanzee's brain."

From this time forth, the untenability of his position might have been as apparent to Professor Owen as it was to every one else ; but, so far from retracting the grave errors into which he had fallen, Professor Owen has persisted in and reiterated them ; first, in a lecture delivered before the Royal Institution on the 19th of March, 1861, which is admitted to have been accurately reproduced in the ' Athenæum ' for the 23rd of the same month, in a letter addressed by Professor Owen to that journal on the 30th of March. The 'Athenæum' report was accompanied by a diagram purporting to represent a Gorilla's brain, but in reality so extraordinary a misrepresentation, that Professor Owen substantially, though not explicitly, withdraws it in the letter in question. In amending this error, however, Professor Owen fell into another of much graver import, as his communication concludes with the following paragraph : "For the true proportion in which the cerebrum covers the cerebellum in the highest Apes, reference should be made to the figure of the undissected brain of the Chimpanzee in my 'Reade's Lecture on the Classification, &c. of the Mammalia,' p. 25, fig. 7, 8vo. 1859."

It would not be credible, if it were not unfortunately true, that this figure, to which the trusting public is referred, without a word of qualification, "for the true proportion in which the cerebrum covers the cerebellum in the highest Apes," is exactly that unacknowledged copy of Schroeder van der Kolk and Vrolik's figure whose utter inaccuracy had been pointed out years before by Gratiolet, and had been brought to Professor Owen's knowledge by myself in the passage of my article in the 'Natural History Review' above quoted.

I drew public attention to this circumstance again in my reply to Professor Owen, published in the 'Athenæum' for April 13th, 1861; but the exploded figure was reproduced once more by Professor Owen, without the slightest allusion to its inaccuracy, in the 'Annals of Natural History' for June, 1861 !

This proved too much for the patience of the original authors of the figure, Messrs. Schroeder van der Kolk and Vrolik, who, in a note addressed to the Academy of Amsterdam, of which they were members, declared themselves to be, though decided opponents of all forms of the doctrine of progressive development, above all things, lovers of truth : and that, therefore, at whatever risk of seeming to lend support to views which they disliked, they felt it their duty to take the first opportunity of publicly repudiating Professor Owen's misuse of their authority.

In this note they frankly admitted the justice of the criticisms of M. Gratiolet, quoted above, and they illustrated, by new and careful figures, the posterior lobe, the posterior cornu, and the hippocampus minor of the Orang. Furthermore, having demonstrated the parts, at one of the sittings of the Academy, they add, " la présence des parties contestées y a été universellement reconnue par les anatomistes présents à la séance. La seul doute qui soit resté se rapporte au pcs Hippocampi minor. . . . . A l'état frais l'indice du petit pied d'Hippocampe était plus prononcé que maintenant."

Professor Owen repeated his erroneous assertions at the meeting of the British Association in 1861, and again, without any obvious necessity, and without adducing a single new fact or new argument, or being able in any way to meet the crushing evidence from original dissections of numerous Apes' brains, which had in the meanwhile been brought forward by Prof. Rolleston,[*] F.R.S., Mr. Marshall,[†] F.R.S., Mr. Flower,[‡] Mr. Turner [§] and myself,[‖] revived the subject at the Cambridge meeting of the same body in 1862. Not content with the tolerably vigorous repudiation which these unprecedented proceedings met with in Section D, Professor Owen sanctioned the publication of a version of his own statements, accompanied by a strange misrepresentation of mine (as may be seen by comparison of the 'Times' Report of the discussion), in the 'Medical Times' for October 11th, 1862. I subjoin the conclusion of my reply in the same journal for October 25th.

" If this were a question of opinion, or a question of interpretation of parts or of terms,—were it even a question of observation in which the testimony of my own senses alone was pitted against that of another person, I should adopt a very different tone in discussing this matter. I should, in all humility, admit the likelihood of having myself erred in judgment, failed in knowledge, or been blinded by prejudice.

" But no one pretends now that the controversy is one of terms or of opinions. Novel and devoid of authority as some of Professor Owen's proposed definitions may have been, they might be accepted without changing the great features of the case. Hence, though special investigations into these matters have been undertaken during the last two years by Dr. Allen Thomson, by Dr. Rolleston,

* On the Affinities of the Brain of the Orang. Nat. Hist. Review, April, 1861.

† On the Brain of a young Chimpanzee. Ibid. July, 1861.

‡ On the Posterior lobes of the Cerebrum of the Quadrumana. Philosophical Transactions, 1862.

§ On the anatomical Relations of the Surfaces of the Tentorium to the Cerebrum and Cerebellum in Man and the lower Mammals. Proceedings of the Royal Society of Edinburgh, March, 1862.

‖ On the Brain of Ateles. Proceedings of Zoological Society, 1861.

by Mr. Marshall, and by Mr. Flower, all, as you are aware, anatomists of repute in this country, and by Professors Schroeder van der Kolk, and Vrolik (whom Professor Owen incautiously tried to press into his own service) on the Continent, all these able and conscientious observers have with one accord testified to the accuracy of my statements, and to the utter baselessness of the assertions of Professor Owen. Even the venerable Rudolph Wagner, whom no man will accuse of progressional proclivities, has raised his voice on the same side; while not a single anatomist, great or small, has supported Professor Owen.

"Now, I do not mean to suggest that scientific differences should be settled by universal suffrage, but I do conceive that solid proofs must be met by something more than empty and unsupported assertions. Yet during the two years through which this preposterous controversy has dragged its weary length, Professor Owen has not ventured to bring forward a single preparation in support of his often-repeated assertions.

"The case stands thus, therefore:—Not only are the statements made by me in consonance with the doctrines of the best older authorities, and with those of all recent investigators, but I am quite ready to demonstrate them on the first monkey that comes to hand; while Professor Owen's assertions are not only in diametrical opposition to both old and new authorities, but he has not produced, and, I will add, cannot produce, a single preparation which justifies them."

I now leave this subject, for the present.—For the credit of my calling I should be glad to be, hereafter, for ever silent upon it. But, unfortunately, this is a matter upon which, after all that has occurred, no mistake or confusion of terms is possible—and in affirming that the posterior lobe, the posterior cornu, and the hippocampus minor exist in certain Apes, I am stating either that which is true, or that which I must know to be false. The question has thus become one of personal veracity. For myself, I will accept no other issue than this, grave as it is, to the present controversy.

III.

ON SOME FOSSIL REMAINS OF MAN.

I HAVE endeavoured to show in the preceding Essay, that the ANTHROPINI, or Man Family, form a very well defined group of the Primates, between which and the immediately following Family, the CATARHINI, there is, in the existing world, the same entire absence of any transitional form or connecting link, as between the CATARHINI and PLATYRHINI.

It is a commonly received doctrine, however, that the structural intervals between the various existing modifications of organic beings may be diminished, or even obliterated, if we take into account the long and varied succession of animals and plants which have preceded those now living and which are known to us only by their fossilized remains. How far this doctrine is well based, how far, on the other hand, as our knowledge at present stands, it is an overstatement of the real facts of the case, and an exaggeration of the conclusions fairly deducible from them, are points of grave importance, but into the discussion of which I do not, at present, propose to enter. It is enough that such a view of the relations of extinct to living beings has been propounded, to lead us to inquire, with anxiety, how far the recent discoveries of human remains in a fossil state bear out, or oppose, that view.

I shall confine myself, in discussing this question, to those fragmentary Human skulls from the caves of Engis in the valley of the Meuse, in Belgium, and of the Neanderthal near Düsseldorf, the geological relations of which have been examined with so much care by Sir Charles Lyell; upon whose high authority I shall take it for granted, that the Engis skull belonged to a contemporary of the Mammoth (*Elephas primigenius*) and of the woolly Rhinoceros (*Rhinocerus tichorhinus*), with the bones of which it was found associated; and that the Neanderthal skull is of great, though uncertain, antiquity. Whatever be the geological age of the latter skull, I con-

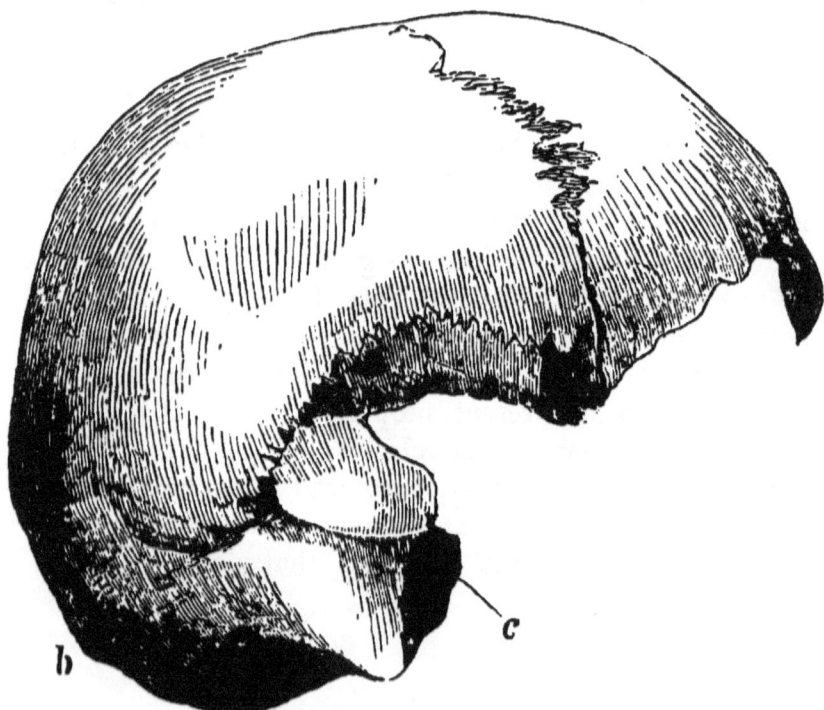

Fig. 23.—The skull from the cave of Engis—viewed from the right side. One half the size of nature. *a* glabella, *b* occipital protuberance. (*a* to *b* glabello-occipital line), *c* auditory foramen.

ceive it is quite safe (on the ordinary principles of paleontological reasoning) to assume that the former takes us to, at least, the further side of the vague biological limit which separates the present geological epoch from that which immediately preceded it. And there can be no doubt that the physical geography of Europe has changed wonderfully since the bones of Men and Mammoths, Hyænas and Rhinoceroses were washed pell-mell into the cave of Engis.

The skull from the cave of Engis was originally discovered by Professor Schmerling, and was described by him, together with other human remains disinterred at the same time, in his valuable work, " Recherches sur les ossemens fossiles découverts dans les cavernes de la Province de Liege," published in 1833, (p. 59, *et seq.*) from which the following paragraphs are extracted, the precise expressions of the author being, as far as possible, preserved.

" In the first place, I must remark that these human remains, which are in my possession, are characterized, like the thousands of bones which I have lately been disinterring, by the extent of the decomposition which they have undergone, which is precisely the same as that of the extinct species : all, with a few exceptions, are broken ; some few are rounded, as is frequently found to be the case in fossil remains of other species. The fractures are vertical or oblique ; none of them are eroded ; their colour does not differ from that of other fossil bones, and varies from whitish yellow to blackish. All are lighter than recent bones, with the exception of those which have a calcareous incrustation, and the cavities of which are filled with such matter.

The cranium which I have caused to be figured, Plate

I, figs. 1, 2, is that of an old person.   The sutures are beginning to be effaced : all the facial bones are wanting, and of the temporal bones only a fragment of that of the right side is preserved.

The face and the base of the cranium had been detached before the skull was deposited in the cave, for we were unable to find those parts, though the whole cavern was regularly searched.   The cranium was met with at a depth of a metre and a half [five feet nearly] hidden under an osseous breccia, composed of the remains of small animals, and containing one rhinoceros tusk, with several teeth of horses and of ruminants.   This breccia, which has been spoken of above, (p. 31) was a metre [3¼ feet about] wide, and rose to the height of a metre and a half above the floor of the cavern, to the walls of which it adhered strongly.

The earth which contained this human skull exhibited no trace of disturbance : teeth of rhinoceros, horse, hyæna, and bear, surrounded it on all sides.

The famous Blumenbach* has directed attention to the differences presented by the form and the dimensions of human crania of different races.   This important work would have assisted us greatly, if the face, a part essential for the determination of race, with more or less accuracy, had not been wanting in our fossil cranium.

We are convinced that even if the skull had been complete, it would not have been possible to pronounce, with certainty, upon a single specimen ; for individual variations are so numerous in the crania of one and the same race, that one cannot, without laying oneself open to large chances of error, draw any inference from a single frag-

* Decas Collectionis suæ craniorum diversarum gentium illustrata. Gottingæ, 1790–1820.

ment of a cranium to the general form of the head to which it belonged.

Nevertheless, in order to neglect no point respecting the form of this fossil skull, we may observe that, from the first, the elongated and narrow form of the forehead attracted our attention.

In fact, the slight elevation of the frontal, its narrowness, and the form of the orbit, approximate it more nearly to the cranium of an Ethiopian than to that of an European : the elongated form and the produced occiput are also characters which we believe to be observable in our fossil cranium ; but to remove all doubt upon that subject I have caused the contours of the cranium of an European and of an Ethiopian to be drawn and the foreheads represented.   Plate II, figs. 1 & 2, and, in the same plate, figs. 3 & 4, will render the differences easily distinguishable ; and a single glance at the figures, will be more instructive than a long and wearisome description.

At whatever conclusion we may arrive as to the origin of the man from whence this fossil skull proceeded, we may express an opinion without exposing ourselves to a fruitless controversy.   Each may adopt the hypothesis which seems to him most probable : for my own part, I hold it to be demonstrated that this cranium has belonged to a person of limited intellectual faculties, and we conclude thence that it belonged to a man of a low degree of civilization : a deduction which is borne out by contrasting the capacity of the frontal with that of the occipital region.

Another cranium of a young individual was discovered in the floor of the cavern beside the tooth of an elephant ; the skull was entire when found, but the moment it was lifted it fell into pieces, which I have not, as yet, been able to put together again.   But I have represented the bones of the upper jaw, Plate I, fig. 5.   The state of the alveoli and the teeth, shows that the molars had not yet pierced the gum. Detached milk molars and some fragments of a human skull, proceed from this same place. The figure 3, represents a human superior incisor tooth, the size of which is truly remarkable.*

Figure 4 is a fragment of a superior maxillary bone, the molar teeth of which are worn down to the roots.

I possess two vertebræ, a first and last dorsal.

A clavicle of the left side (see Plate III, fig. 1); although it belonged to a young individual, this bone shows that he must have been of great stature.†

Two fragments of the radius, badly preserved, do not indicate that the height of the man, to whom they belonged, exceeded five feet and a half.

As to the remains of the upper extremities, those which are in my possession, consist merely of a fragment of an ulna and of a radius (Plate III, fig. 5 and 6).

Figure 2, Plate IV, represents a metacarpal bone, contained in the breccia, of which we have spoken; it was found in the lower part above the cranium : add to this some metacarpal bones, found at very different distances, half-a-dozen metatarsals, three phalanges of the hand, and one of the foot.

This is a brief enumeration of the remains of human bones collected in the cavern of Engis, which has preserved for us the remains of three individuals, surrounded by those of the Elephant, of the Rhinoceros, and of Carnivora of species unknown in the present creation."

---

* In a subsequent passage, Schmerling remarks upon the occurrence of an incisor tooth ' of enormous size' from the caverns of Engihoul. The tooth figured is somewhat long, but its dimensions do not appear to me to be otherwise remarkable.

† The figure of this clavicle measures 5 inches from end to end in a straight line—so that the bone is rather a small than a large one.

From the cave of Engihoul, opposite that o. Engis, on the right bank of the Meuse, Schmerling obtained the remains of three other individuals of Man, among which were only two fragments of parietal bones, but many bones of the extremities. In one case, a broken fragment of an ulna was soldered to a like fragment of a radius by stalagmite, a condition frequently observed among the bones of the Cave Bear (*Ursus spelæus*), found in the Belgian caverns.

It was in the cavern of Engis that Professor Schmerling found, incrusted with stalagmite and joined to a stone, the pointed bone implement, which he has figured in fig. 7 of his Plate XXXVI, and worked flints were found by him in all those Belgian caves, which contained an abundance of fossil bones.

A short letter from M. Geoffroy St. Hilaire, published in the Comptes Rendus of the Academy of Sciences of Paris, for July 2nd, 1838, speaks of a visit (and apparently a very hasty one) paid to the collection of Professor 'Schermidt' (which is presumably a misprint for Schmerling) at Liège. The writer briefly criticises the drawings which illustrate Schmerling's work, and affirms that the " human cranium is a little longer than it is represented " in Schmerling's figure. The only other remark worth quoting is this :—" The aspect of the human bones differs little from that of the cave bones, with which we are familiar, and of which there is a considerable collection in the same place. With respect to their special forms, compared with those of the varieties of recent human crania, few *certain* conclusions can be put forward; for much greater differences exist between the different specimens of well-characterized varieties, than between the fossil cranium of Liège and that of one of those varieties selected as a term of comparison."

7

Geoffroy St. Hilaire's remarks are, it will be observed, little but an echo of the philosophic doubts of the describer and discoverer of the remains. As to the critique upon Schmerling's figures, I find that the side view given by the latter is really about $\frac{3}{10}$ ths of an inch shorter than the original, and that the front view is diminished to about the same extent. Otherwise the representation is not, in any way, inaccurate, but corresponds very well with the cast which is in my possession.

A piece of the occipital bone, which Schmerling seems to have missed, has since been fitted on to the rest of the cranium by an accomplished anatomist, Dr. Spring of Liège, under whose direction an excellent plaster cast was made for Sir Charles Lyell. It is upon and from a duplicate of that cast that my own observations and the accompanying figures, the outlines of which are copied from the very accurate Camera lucida drawings, by my friend Mr. Busk, reduced to one-half of the natural size, are made.

As Professor Schmerling observes, the base of the skull is destroyed, and the facial bones are entirely absent; but the roof of the cranium, consisting of the frontal, parietal, and the greater part of the occipital bones, as far as the middle of the occipital foramen, is entire or nearly so. The left temporal bone is wanting. Of the right temporal, the parts in the immediate neighbourhood of the auditory foramen, the mastoid process, and a considerable portion of the squamous element of the temporal are well preserved (Fig 23).

The lines of fracture which remain between the coadjusted pieces of the skull, and are faithfully displayed in Schmerling's figure, are readily traceable in the cast. The sutures are also discernible, but the complex disposition of their serrations, shown in the figure, is not obvious in the cast. Though the ridges which give attachment to

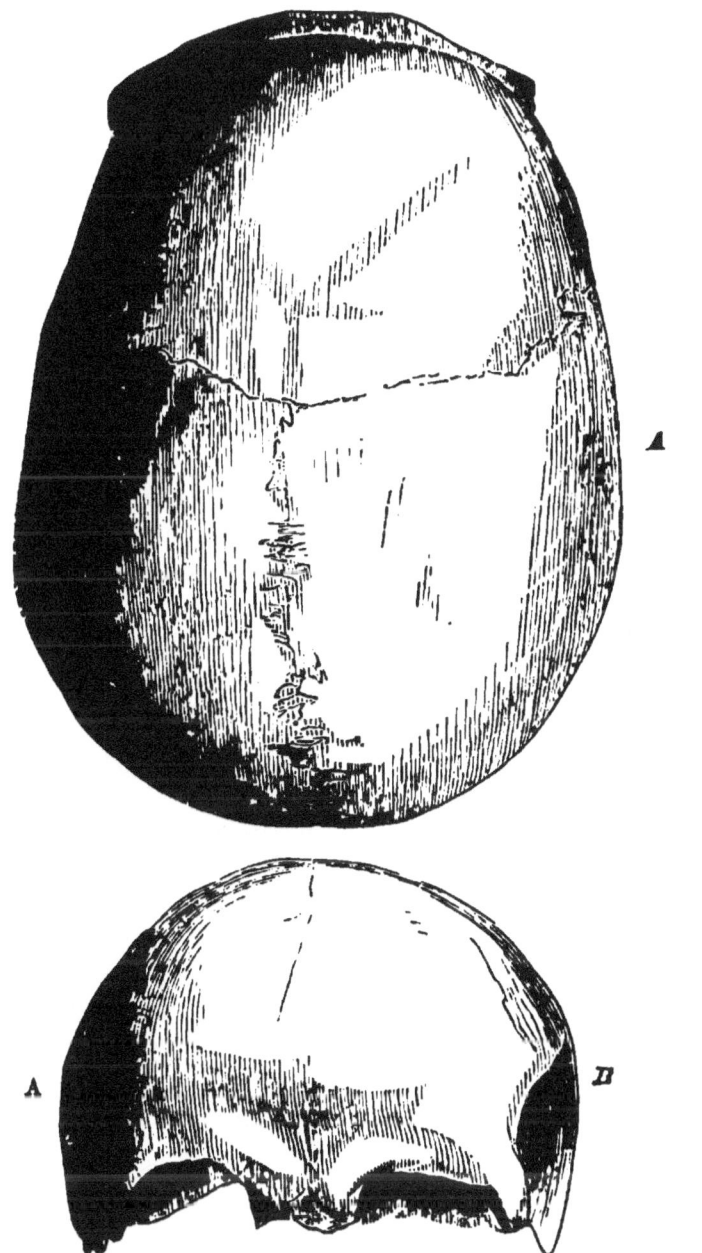

FIG. 24.—The Engis skull viewed from above (A) and in front (B).

muscles are not excessively prominent, they are well marked, and taken together with the apparently well developed frontal sinuses, and the condition of the sutures, leave no doubt on my mind that the skull is that of an adult, if not middle-aged man.

The extreme length of the skull is 7.7 inches. Its extreme breadth, which corresponds very nearly with the interval between the parietal protuberances, is not more than 5.4 inches. The proportion of the length to the breadth is therefore very nearly as 100 to 70. If a line be drawn from the point at which the brow curves in towards the root of the nose, and which is called the 'glabella' (a), (fig. 23), to the occipital protuberance (b), and the distance to the highest point of the arch of the skull be measured perpendicularly from this line, it will be found to be 4.75 inches. Viewed from above, fig. 24 A, the forehead presents an evenly rounded curve, and passes into the contour of the sides and back of the skull, which describes a tolerably regular elliptical curve.

The front view (fig. 24 B) shows that the roof of the skull was very regularly and elegantly arched in the transverse direction, and that the transverse diameter was a little less below the parietal protuberances, than above them. The forehead cannot be called narrow in relation to the rest of the skull, nor can it be called a retreating forehead; on the contrary, the antero-posterior contour of the skull is well arched, so that the distance along that contour, from the nasal depression to the occipital protuberance, measures about 13.75 inches. The transverse arc of the skull, measured from one auditory foramen to the other, across the middle of the sagittal suture, is about 13 inches. The sagittal suture itself is 5.5 inches long.

The supraciliary prominences or brow-ridges (on each side of a, fig. 23) are well, but not excessively, developed, and are separated by a median depression. Their principal elevation is disposed so obliquely that I judge them to be due to large frontal sinuses.

If a line joining the glabella and the occipital protuberance (a, b, fig. 23) be made horizontal, no part of the occipital region projects more than $\frac{1}{15}$th of an inch behind the posterior extremity of that line, and the upper edge of the auditory foramen (c) is almost in contact with a line drawn parallel with this upon the outer surface of the skull.

A transverse line drawn from one auditory foramen to the other traverses, as usual, the forepart of the occipital foramen. The capacity of the interior of this fragmentary skull has not been ascertained.

The history of the Human remains from the cavern in the Neanderthal may best be given in the words of their original describer, Dr. Schaaffhausen,* as translated by Mr. Busk.

"In the early part of the year 1857, a human skeleton was discovered in a limestone cave in the Neanderthal, near Hochdal, between Düsseldorf and Elberfeld. Of this, however, I was unable to procure more than a plaster cast of the cranium, taken at Elberfeld, from which I drew up an account of its remarkable conformation, which was, in the first instance, read on the 4th of February, 1857, at the meeting of the Lower Rhine Medical and Natural History Society, at Bonn.† Subsequently

* ON THE CRANIA OF THE MOST ANCIENT RACES OF MAN. By Professor D. Schaaffhausen, of Bonn. (From Müller's Archiv., 1858, pp. 453.) With Remarks, and original Figures, taken from a Cast of the Neanderthal Cranium. By George Busk, F.R.S., &c. Natural History Review, April, 1861.

† Verhandl. d. Naturhist. Vereins der preuss. Rheinlande und Westphalens., xiv. Bonn, 1857.

Dr. Fuhlrott, to whom science is indebted for the preservation of these bones, which were not at first regarded as human, and into whose possession they afterwards came, brought the cranium from Elberfeld to Bonn, and entrusted it to me for a more accurate anatomical examination. At the General Meeting of the Natural History Society of Prussian Rhineland and Westphalia, at Bonn, on the 2nd of June, 1857,* Dr. Fuhlrott himself gave a full account of the locality and of the circumstances under which the discovery was made. He was of opinion that the bones might be regarded as fossil; and in coming to this conclusion, he laid especial stress upon the existence of dendritic deposits, with which their surface was covered, and which were first noticed upon them by Professor Mayer. To this communication I appended a brief report on the results of my anatomical examination of the bones. The conclusions at which I arrived were :—1st. That the extraordinary form of the skull was due to a natural conformation hitherto not known to exist, even in the most barbarous races. 2nd. That these remarkable human remains belonged to a period antecedent to the time of the Celts and Germans, and were in all probability derived from one of the wild races of North-western Europe, spoken of by Latin writers ; and which were encountered as autochthones by the German immigrants. And 3rdly. That it was beyond doubt that these human relics were traceable to a period at which the latest animals of the diluvium still existed ; but that no proof of this assumption, nor consequently of their so-termed *fossil* condition, was afforded by the circumstances under which the bones were discovered.

As Dr. Fuhlrott has not yet published his description of these circumstances, I borrow the following account of

* Ib. Correspondenzblatt. No. 2.

them from one of his letters. "A small cave or grotto, high enough to admit a man, and about 15 feet deep from the entrance, which is 7 or 8 feet wide, exists in the southern wall of the gorge of the Neanderthal, as it is termed, at a distance of about 10 feet from the Düssel, and about 60 feet above the bottom of the valley. In its earlier and uninjured condition, this cavern opened upon a narrow plateau lying in front of it, and from which the rocky wall descended almost perpendicularly into the river. It could be reached, though with difficulty, from above. The uneven floor was covered to a thickness of 4 or 5 feet with a deposit of mud, sparingly intermixed with rounded fragments of chert. In the removing of this deposit, the bones were discovered. The skull was first noticed, placed nearest to the entrance of the cavern ; and further in, the other bones, lying in the same horizontal plane. Of this I was assured in the most positive terms, by two labourers who were employed to clear out the grotto, and who were questioned by me on the spot. At first no idea was entertained of the bones being human ; and it was not till several weeks after their discovery that they were recognised as such by me, and placed in security. But, as the importance of the discovery was not at the time perceived, the labourers were very careless in the collecting, and secured chiefly only the larger bones ; and to this circumstance it may be attributed that fragments merely of the probably perfect skeleton came into my possession."

"My anatomical examination of these bones afforded the following results :—

The cranium is of unusual size, and of a long-elliptical form. A most remarkable peculiarity is at once obvious in the extraordinary development of the frontal sinuses, owing to which the superciliary ridges, which coalesce completely in the middle, are rendered so prominent, that the frontal bone exhibits a considerable hollow or depression above or rather behind them, whilst a deep depression is also formed in the situation of the root of the nose. The forehead is narrow and low, though the middle and hinder portions of the cranial arch are well developed. Unfortunately, the fragment of the skull that has been preserved consists only of the portion situated above the roof of the orbits and the superior occipital ridges, which are greatly developed, and almost conjoined so as to form a horizontal eminence. It includes almost the whole of the frontal bone, both parietals, a small part of the squamous and the upper-third of the occipital. The recently fractured surfaces show that the skull was broken at the time of its disinterment. The cavity holds 16,876 grains of water, whence its cubical contents may be estimated at 57.64 inches, or 1033.24 cubic centimetres. In making this estimation, the water is supposed to stand on a level with the orbital plate of the frontal, with the deepest notch in the squamous margin of the parietal, and with the superior semicircular ridges of the occipital. Estimated in dried millet-seed, the contents equalled 31 ounces, Prussian Apothecaries' weight. The semicircular line indicating the upper boundary of the attachment of the temporal muscle, though not very strongly marked, ascends nevertheless to more than half the height of the parietal bone. On the right superciliary ridge is observable an oblique furrow or depression, indicative of an injury received during life.* The coronal and sagittal sutures are on the exterior nearly closed, and on the inside so completely ossified as to have left no traces whatever, whilst the lambdoidal remains quite open. The depressions for the Pacchionian glands are deep and numerous;

---

* This, Mr. Busk has pointed out, is probably the notch for the frontal nerve.

and there is an unusually deep vascular groove immediately behind the coronal suture, which, as it terminates in a foramen, no doubt transmitted a *vena emissaria*. The course of the frontal suture is indicated externally by a slight ridge; and where it joins the coronal, this ridge rises into a small protuberance. The course of the sagittal suture is grooved, and above the angle of the occipital bone the parietals are depressed.

mm*

The length of the skull from the nasal process of the frontal over the vertex to the superior semicircular lines of the occipital measures . . . . 303 (300) = 12·0".

Circumference over the orbital ridges and the superior semicircular lines of the occipital . 590 (590)=23·37" or 23".

Width of the frontal from the middle of the temporal line on one side to the same point on the opposite . . . . . 104 (114) = 4·1"—4·5".

Length of the frontal from the nasal process to the coronal suture . . . . 133 (125) = 5·25"—5".

Extreme width of the frontal sinuses . . . . 25 (23) = 1·0" —0·9".

Vertical height above a line joining the deepest notches in the squamous border of the parietals . . . . 70 = 2·75."

Width of hinder part of skull from one parietal protuberance to the other . . . . . . 138 (150) = 5·4" —5·9".

* The numbers in brackets are those which I should assign to the different measures, as taken from the plaster cast.—G. B.

Distance from the upper angle
of the occipital to the supe-
rior semicircular lines . . . 51 (60) = 1·9″ — 2·4″.
Thickness of the bone at the
parietal protuberance . . . 8.
—— at the angle of the occipital 9.
—— at the superior semicircular
line of the occipital . . . . 10 = 0·3″.

Besides the cranium, the following bones have been
secured :—

1. Both thigh-bones, perfect. These, like the skull,
and all the other bones, are characterized by their unusual
thickness, and the great development of all the elevations
and depressions for the attachment of muscles. In the
Anatomical Museum at Bonn, under the designation of
"Giant's-bones," are some recent thigh-bones, with which
in thickness the foregoing pretty nearly correspond, although they are shorter.

	Giant's bones. mm.	Fossil bones. mm.
Length	542 = 21·4″	438 = 17·4″.
Diameter of head of femur.	54 = 2·14″	53 = 2·0″.
" of lower articular end, from one condyle to the other	89 = 3·5″	87 = 3·4″.
Diameter of femur in the middle	33 = 1·2″	30 = 1·1″.

2. A perfect right humerus, whose size shows that it
belongs to the thigh-bones.

	mm.
Length	312 = 12·3″.
Thickness in the middle . .	26 = 1·0″
Diameter of head	49 = 1·9″.

Also a perfect right radius of corresponding dimensions, and the upper-third of a right ulna corresponding to the humerus and radius.

3. A left humerus, of which the upper-third is wanting, and which is so much slenderer than the right as apparently to belong to a distinct individual; a left *ulna*, which, though complete, is pathologically deformed, the coronoid process being so much enlarged by bony growth, that flexure of the elbow beyond a right angle must have been impossible; the anterior fossa of the humerus for the reception of the coronoid process being also filled up with a similar bony growth. At the same time, the olecranon is curved strongly downwards. As the bone presents no sign of rachitic degeneration, it may be supposed that an injury sustained during life was the cause of the anchylosis. When the left ulna is compared with the right radius, it might at first sight be concluded that the bones respectively belonged to different individuals, the ulna being more than half an inch too short for articulation with a corresponding radius. But it is clear that this shortening, as well as the attenuation of the left humerus, are both consequent upon the pathological condition above described.

4. A left *ilium*, almost perfect, and belonging to the femur; a fragment of the right *scapula;* the anterior extremity of a rib of the right side; and the same part of a rib of the left side; the hinder part of a rib of the right side; and, lastly, two hinder portions and one middle portion of ribs, which, from their unusually rounded shape, and abrupt curvature, more resemble the ribs of a carnivorous animal than those of a man. Dr. H. v. Meyer, however, to whose judgment I defer, will not venture to declare them to be ribs of any animal; and it only remains to suppose that this abnormal condition has arisen from

an unusually powerful development of the thoracic muscles.

The bones adhere strongly to the tongue, although, as proved by the use of hydrochloric acid, the greater part of the cartilage is still retained in them, which appears, however, to have undergone that transformation into gelatine which has been observed by v. Bibra in fossil bones. The surface of all the bones is in many spots covered with minute black specks, which, more especially under a lens, are seen to be formed of very delicate *dendrites*. These deposits, which were first observed on the bones by Dr. Mayer, are most distinct on the inner surface of the cranial bones. They consist of a ferruginous compound, and, from their black colour, may be supposed to contain manganese. Similar dendritic formations also occur, not unfrequently, on laminated rocks, and are usually found in minute fissures and cracks. At the meeting of the Lower Rhine Society at Bonn, on the 1st April, 1857, Prof. Mayer stated that he had noticed in the museum of Poppelsdorf similar dendritic crystallizations on several fossil bones of animals, and particularly on those of *Ursus spelœus*, but still more abundantly and beautifully displayed on the fossil bones and teeth of *Equus adamiticus, Elephas primigenius,* &c., from the caves of Bolve and Sundwig. Faint indications of similar *dendrites* were visible in a Roman skull from Siegburg; whilst other ancient skulls, which had lain for centuries in the earth, presented no trace of them.* I am indebted to H. v. Meyer for the following remarks on this subject :—

" The incipient formation of dendritic deposits, which were formerly regarded as a sign of a truly fossil condition, is interesting. It has even been supposed that in diluvial deposits the presence of *dendrites* might be re-

* Verh. des Naturhist. Vereins in Bonn, xiv. 1857.

garded as affording a certain mark of distinction between bones mixed with the diluvium at a somewhat later period and the true diluvial relics, to which alone it was supposed that these deposits were confined. But I have long been convinced that neither can the absence of *dendrites* be regarded as indicative of recent age, nor their presence as sufficient to establish the great antiquity of the objects upon which they occur. I have myself noticed upon paper, which could scarcely be more than a year old, dendritic deposits, which could not be distinguished from those on fossil bones. Thus I possess a dog's skull from the Roman colony of the neighbouring Heddersheim, *Castrum Hadrianum*, which is in no way distinguishable from the fossil bones from the Frankish caves ; it presents the same colour, and adheres to the tongue just as they do ; so that this character also, which, at a former meeting of German naturalists at Bonn, gave rise to amusing scenes between Buckland and Schmerling, is no longer of any value. In disputed cases, therefore, the condition of the bone can scarcely afford the means for determining with certainty whether it be fossil, that is to say, whether it belong to geological antiquity or to the historical period."

As we cannot now look upon the primitive world as representing a wholly different condition of things, from which no transition exists to the organic life of the present time, the designation of *fossil* as applied to *a bone*, has no longer the sense it conveyed in the time of Cuvier. Sufficient grounds exist for the assumption that man coexisted with the animals found in the *diluvium ;* and many a barbarous race may, before all historical time, have disappeared together with the animals of the ancient world, whilst the races whose organization is improved have continued the genus. The bones which form the subject of

this paper present characters which, although not decisive
as regards a geological epoch, are, nevertheless, such as
indicate a very high antiquity. It may also be remarked
that, common as is the occurrence of diluvial animal bones
in the muddy deposits of caverns, such remains have not
hitherto been met with in the caves of the Neanderthal;
and that the bones, which were covered by a deposit of
mud not more than four or five feet thick, and without any
protective covering of stalagmite, have retained the greatest part of their organic substance.

These circumstances might be adduced against the
probability of a geological antiquity. Nor should we be
justified in regarding the cranial conformation as perhaps
representing the most savage primitive type of the human
race, since crania exist among living savages, which,
though not exhibiting such a remarkable conformation of
the forehead, which gives the skull somewhat the aspect
of that of the large apes, still in other respects, as for instance in the greater depth of the temporal fossæ, the
crest-like, prominent temporal ridges, and a generally less
capacious cranial cavity, exhibit an equally low stage of
development. There is no reason for supposing that the
deep frontal hollow is due to any artificial flattening, such
as is practised in various modes by barbarous nations in
the Old and New World. The skull is quite symmetrical,
and shows no indication of counter-pressure at the occiput,
whilst, according to Morton, in the Flat-heads of the Co-
lumbia, the frontal and parietal bones are always unsym-
metrical. Its conformation exhibits the sparing development of the anterior part of the head which has been so
often observed in very ancient crania, and affords one of
the most striking proofs of the influence of culture and
civilization on the form of the human skull."

In a subsequent passage, Dr. Schaaffhausen remarks:

" There is no reason whatever for regarding the unusual development of the frontal sinuses in the remarkable
skull from the Neanderthal as an individual or pathological deformity; it is unquestionably a typical race-charac-
ter, and is physiologically connected with the uncommon
thickness of the other bones of the skeleton, which exceeds
by about one-half the usual proportions.   This expansion
of the frontal sinuses, which are appendages of the air-
passages, also indicates an unusual force and power of en-
durance in the movements of the body, as may be concluded from the size of all the ridges and processes for the
attachment of the muscles or bones.   That this conclusion
may be drawn from the existence of large frontal sinuses,
and a prominence of the lower frontal region, is confirmed
in many ways by other observations.   By the same characters, according to Pallas, the wild horse is distinguished
from the domesticated, and, according to Cuvier, the fossil
cave-bear from every recent species of bear, whilst, according to Roulin, the pig, which has become wild in America,
and regained a resemblance to the wild boar, is thus distinguished from the same animal in the domesticated
state, as is the chamois from the goat ; and, lastly, the
bull-dog, which is characterised by its large bones and
strongly-developed muscles from every other kind of dog.
The estimation of the facial angle, the determination of
which, according to Professor Owen, is also difficult in the
great apes, owing to the very prominent supra-orbital
ridges, in the present case is rendered still more difficult
from the absence both of the auditory opening and of the
nasal spine.   But if the proper horizontal position of the
skull be taken from the remaining portions of the orbital
plates, and the ascending line made to touch the surface
of the frontal bone behind the prominent supra-orbital

ridges, the facial angle is not found to exceed 56°.* Unfortunately, no portions of the facial bones, whose conformation is so decisive as regards the form and expression of the head, have been preserved. The cranial capacity, compared with the uncommon strength of the corporeal frame, would seem to indicate a small cerebral development. The skull, as it is, holds about 31 ounces of millet-seed; and as, from the proportionate size of the wanting bones, the whole cranial cavity should have about 6 ounces more added, the contents, were it perfect, may be taken at 37 ounces. Tiedemann assigns, as the cranial contents in the Negro, 40, 38, and 35 ounces. The cranium holds rather more than 36 ounces of water, which corresponds to a capacity of 1033·24 cubic centimetres. Huschke estimates the cranial contents of a Negress at 1127 cubic centimetres; of an old Negro at 1146 cubic centimetres. The capacity of the Malay skulls, estimated by water, equalled 36, 33 ounces, whilst in the diminutive Hindoos it falls to as little as 27 ounces."

After comparing the Neanderthal cranium with many others, ancient and modern, Professor Schaaffhausen concludes thus :—

" But the human bones and cranium from the Neanderthal exceed all the rest in those peculiarities of conformation which lead to the conclusion of their belonging to a barbarous and savage race. Whether the cavern in which they were found, unaccompanied with any trace of human art, were the place of their interment, or whether, like the bones of extinct animals elsewhere, they had been washed into it, they may still be regarded as the most ancient memorial of the early inhabitants of Europe."

Mr. Busk, the translator of Dr. Schaaffhausen's paper,

* Estimating the facial angle in the way suggested, on the cast I should place it at 64° to 67°.—G. B.

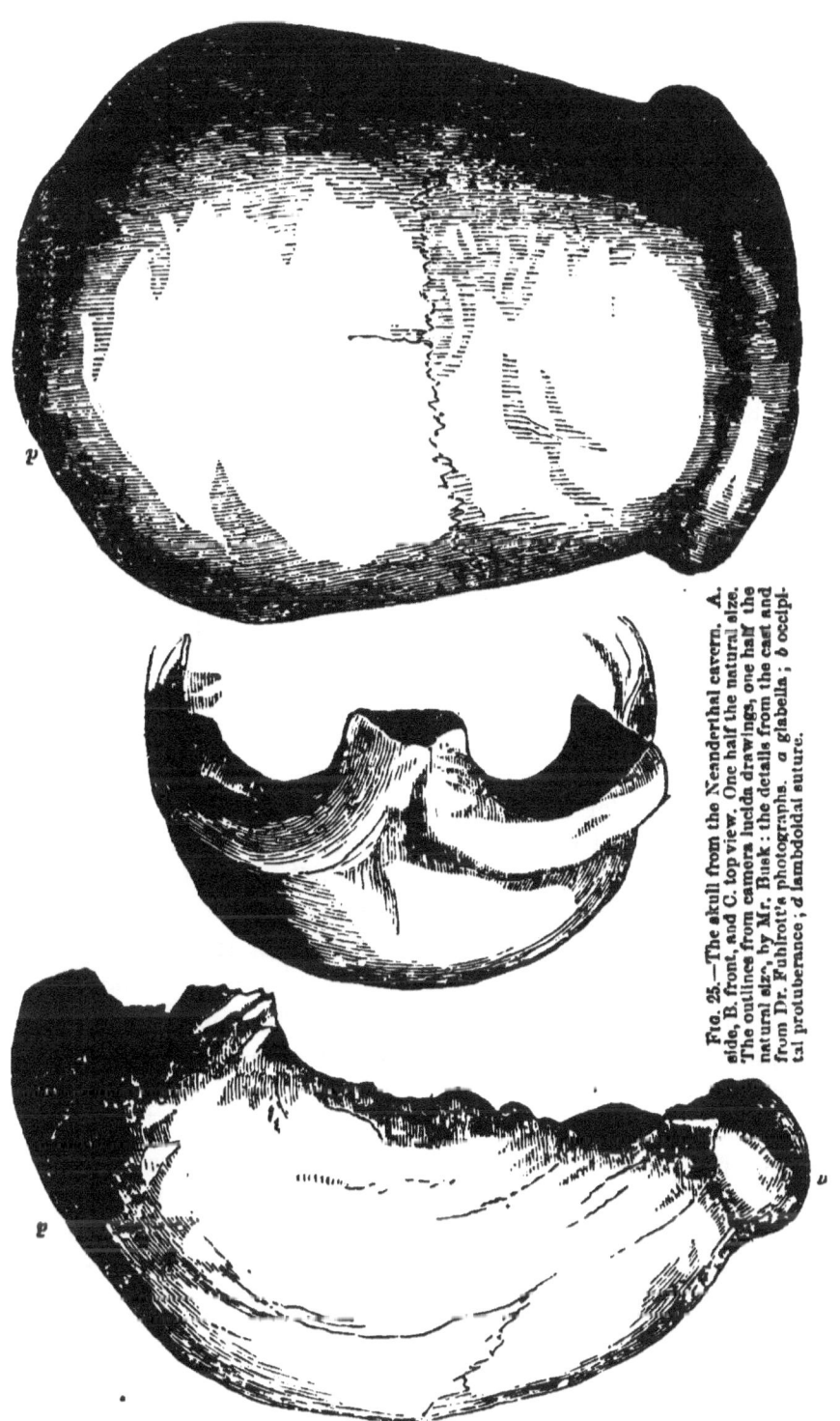

Fig. 25.—The skull from the Neanderthal cavern. A. side, B. front, and C. top view. One half the natural size. The outlines from camera lucida drawings, one half the natural sizr, by Mr. Busk: the details from the cast and from Dr. Fuhlrott's photographs. *a* glabella; *b* occipital protuberance; *d* lambdoidal suture.

has enabled us to form a very vivid conception of the degraded character of the Neanderthal skull, by placing
side by side with its outline, that of the skull of a Chimpanzee, drawn to the same absolute size.

Some time after the publication of the translation of
Professor Schaaffhausen's Memoir, I was led to study the
cast of the Neanderthal cranium with more attention than
I had previously bestowed upon it, in consequence of
wishing to supply Sir Charles Lyell with a diagram, exhibiting the special peculiarities of this skull, as compared
with other human skulls. In order to do this it was
necessary to identify, with precision, those points in the
skulls compared which corresponded anatomically. Of
these points, the glabella was obvious enough ; but when
I had distinguished another, defined by the occipital protuberance and superior semicircular line, and had placed
the outline of the Neanderthal skull against that of the
Engis skull, in such a position that the glabella and occipital protuberance of both were intersected by the same
straight line, the difference was so vast and the flattening
of the Neanderthal skull so prodigious (compare Figs. 23
and 25 A), that I at first imagined I must have fallen into
some error. And I was the more inclined to suspect this,
as, in ordinary human skulls, the occipital protuberance
and superior semicircular curved line on the exterior of
the occiput correspond pretty closely with the 'lateral
sinuses' and the line of attachment of the tentorium inter-
nally. But on the tentorium rests, as I have said in the
preceding Essay, the posterior lobe of the brain ; and
hence, the occipital protuberance, and the curved line in
question, indicate, approximately, the lower limits of that
lobe. Was it possible for a human being to have the brain
thus flattened and depressed ; or, on the other hand, had
the muscular ridges shifted their position ? In order to

solve these doubts, and to decide the question whether the great supraciliary projections did, or did not, arise from the development of the frontal sinuses, I requested Sir Charles Lyell to be so good as to obtain for me from Dr. Fuhlrott, the possessor of the skull, answers to certain queries, and if possible a cast, or at any rate drawings, or photographs, of the interior of the skull.

Dr. Fuhlrott replied, with a courtesy and readiness for

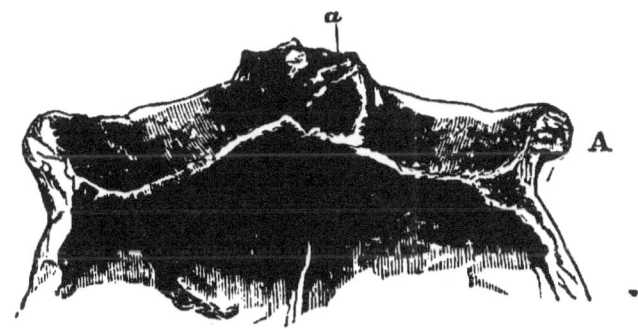

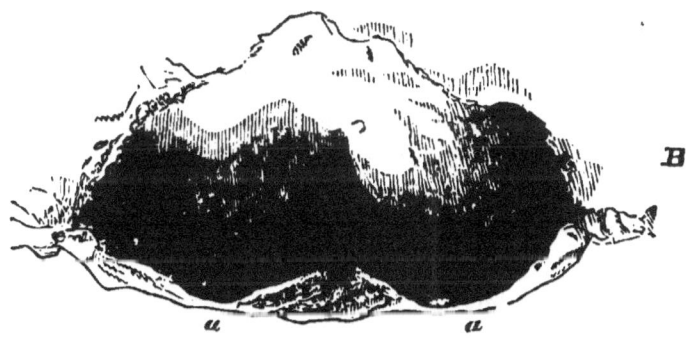

Fig. 26.—Drawings from Dr. Fuhlrott's photographs of parts of the interior of the Neanderthal cranium. A. view of the under and inner surface of the frontal region, showing the inferior apertures of the frontal sinuses (a). B. corresponding view of the occipital region of the skull, showing the impressions of the lateral sinuses (aa).

which I am infinitely indebted to him, to my inquiries, and furthermore sent three excellent photographs. One of these gives a side view of the skull, and from it Fig. 25 A. has been shaded. The second (Fig. 26 A.) exhibits the wide openings of the frontal sinuses upon the inferior surface of the frontal part of the skull, into which, Dr. Fuhlrott writes, " a probe may be introduced to the depth of an inch," and demonstrates the great extension of the thickened supraciliary ridges beyond the cerebral cavity. The third, lastly, (Fig. 26 B.) exhibits the edge and the interior of the posterior, or occipital, part of the skull, and shows very clearly the two depressions for the lateral sinuses, sweeping inwards towards the middle line of the roof of the skull, to form the longitudinal sinus. It was clear, therefore, that I had not erred in my interpretation, and that the posterior lobe of the brain of the Neanderthal man must have been as much flattened as I suspected it to be.

In truth, the Neanderthal cranium has most extraordinary characters. It has an extreme length of 8 inches, while its breath is only 5·75 inches, or, in other words, its length is to its breadth as 100 : 72. It is exceedingly depressed, measuring only about 3·4 inches from the glabello-occipital line to the vertex. The longitudinal arc, measured in the same way as in the Engis skull, is 12 inches; the transverse arc cannot be exactly ascertained, in consequence of the absence of the temporal bones, but was probably about the same, and certainly exceeded 10½ inches. The horizontal circumference is 23 inches. But this great circumference arises largely from the vast development of the supraciliary ridges, though the perimeter of the brain case itself is not small. The large supraciliary ridges give the forehead a far more retreating appearance than its internal contour would bear out.

To an anatomical eye the posterior part of the skull is even more striking than the anterior. The occipital protuberance occupies the extreme posterior end of the skull, when the glabello-occipital line is made horizontal, and so far from any part of the occipital region extending beyond it, this region of the skull slopes obliquely upward and forward, so that the lambdoidal suture is situated well upon the upper surface of the cranium. At the same time, notwithstanding the great length of the skull, the sagittal suture is remarkably short (4½ inches), and the squamosal suture is very straight.

In reply to my questions Dr. Fuhlrott writes that the occipital bone " is in a state of perfect preservation as far as the upper semicircular line, which is a very strong ridge, linear at its extremities, but enlarging towards the middle, where it forms two ridges (bourrelets), united by a linear continuation, which is slightly depressed in the middle."

" Below the left ridge the bone exhibits an obliquely inclined surface, six lines (French) long, and twelve lines wide."

This last must be the surface, the contour of which is shown in Fig. 25 A, below b. It is particularly interesting, as it suggests that, notwithstanding the flattened condition of the occiput, the posterior cerebral lobes must have projected considerably beyond the cerebellum, and as it constitutes one among several points of similarity between the Neanderthal cranium and certain Australian skulls.

Such are the two best known forms of human cranium, which have been found in what may be fairly termed a fossil state. Can either be shown to fill up or diminish, to any appreciable extent, the structural interval which exists between Man and the man-like apes ? Or, on the other hand, does neither depart more widely from the average structure of the human cranium, than normally formed skulls of men are known to do at the present day ?

It is impossible to form any opinion on these questions, without some preliminary acquaintance with the range of variation exhibited by human structure in general—a subject which has been but imperfectly studied, while even of what is known, my limits will necessarily allow me to give only a very imperfect sketch.

The student of anatomy is perfectly well aware that there is not a single organ of the human body the structure of which does not vary, to a greater or less extent, in different individuals. The skeleton varies in the proportions, and even to a certain extent in the connexions, of its constituent bones. The muscles which move the bones vary largely in their attachments. The varieties in the mode of distribution of the arteries are carefully classified, on account of the practical importance of a knowledge of their shiftings to the surgeon. The characters of the brain vary immensely, nothing being less constant than the form and size of the cerebral hemispheres, and the richness of the convolutions upon their surface, while the most changeable structures of all in the human brain, are exactly those on which the unwise attempt has been made to base the distinctive characters of humanity, viz. the posterior cornu of the lateral ventricle, the hippocampus minor, and the degree of projection of the posterior lobe beyond the cerebellum. Finally, as all the world knows, the hair and skin of human beings may present the most extraordinary diversities in colour and in texture.

So far as our present knowledge goes, the majority of the structural varieties to which allusion is here made, are individual. The ape-like arrangement of certain muscles which is occasionally met with* in the white races of mankind, is not known to be more common among Negroes or Australians : nor because the brain of the Hottentot Venus was found to be smoother, to have its convolutions more symmetrically disposed, and to be, so far, more ape-like than that of ordinary Europeans, are we justified in concluding a like condition of the brain to prevail universally among the lower races of mankind, however probable that conclusion may be.

We are, in fact, sadly wanting in information respecting the disposition of the soft and destructible organs of every Race of Mankind but our own ; and even of the skeleton, our Museums are lamentably deficient in every part but the cranium. Skulls enough there are, and since the time when Blumenbach and Camper first called attention to the marked and singular differences which they exhibit, skull collecting and skull measuring has been a zealously pursued branch of Natural History, and the results obtained have been arranged and classified by various writers, among whom the late active and able Retzius must always be the first named.

Human skulls have been found to differ from one another, not merely in their absolute size and in the absolute capacity of the brain case, but in the proportions which the diameters of the latter bear to one another ; in the relative size of the bones of the face (and more particularly of the jaws and teeth) as compared with those of the skull ; in the degree to which the upper jaw (which is of course followed by the lower) is thrown backwards and downwards under the forepart of the brain case, or forwards and upwards in front of and beyond it. They differ further in the relations of the transverse diameter of the face,

---

* See an excellent Essay by Mr. Church on the Myology of the Orang, in the Natural History Review, for 1861.

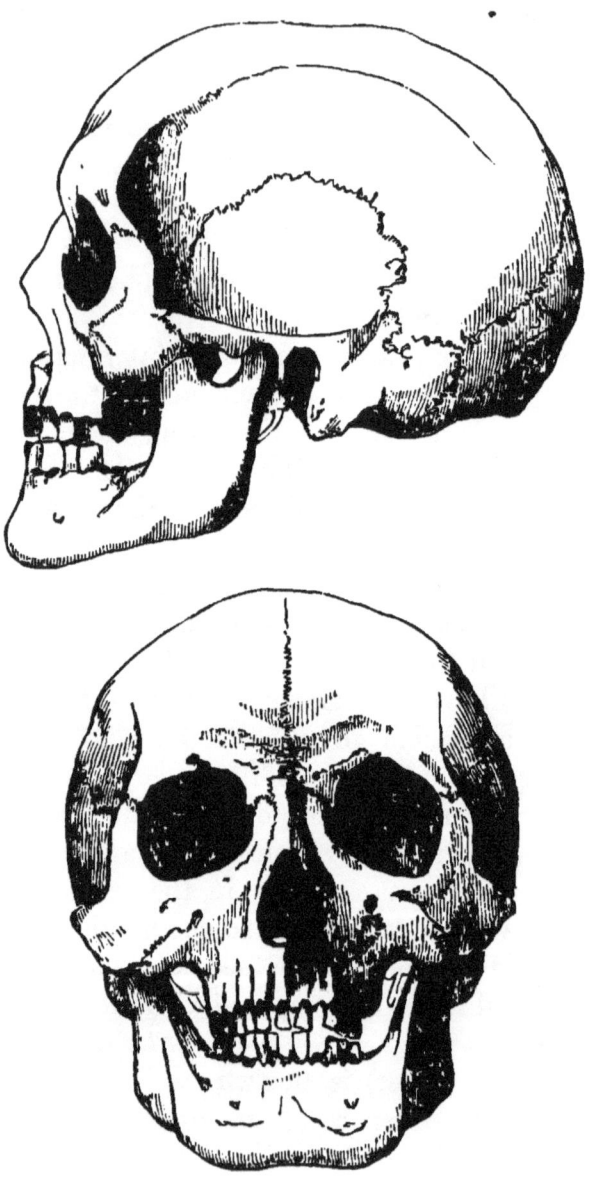

FIG. 27.—Side and front views of the round and orthognathous skull of a
Calmuck after Von Baer.   One-third the natural size.

taken through the cheek bones, to the transverse diameter of the skull; in the more rounded or more gable-like form of the roof of the skull, and in the degree to which the hinder part of the skull is flattened or projects beyond the ridge, into and below which, the muscles of the neck are inserted.

In some skulls the brain case may be said to be ' round,' the extreme length not exceeding the extreme breadth by a greater proportion than 100 to 80, while the difference may be much less.* Men possessing such skulls were termed by Retzius ' *brachycephalic*,' and the skull of a Calmuck, of which a front and side view (reduced outline copies of which are given in figure 27) are depicted by Von Baer in his excellent "Crania selecta," affords a very admirable example of that kind of skull. Other skulls, such as that of a Negro copied in fig. 28 from Mr. Busk's ' Crania typica,' have a very different, greatly elongated form, and may be termed '*oblong*.' In this skull the extreme length is to the extreme breadth as 100 to not more than 67, and the transverse diameter of the human skull may fall below even this proportion. People having such skulls were called by Retzius ' *dolicho-cephalic*.'

The most cursory glance at the side views of these two skulls will suffice to prove that they differ, in another respect, to a very striking extent. The profile of the face of the Calmuck is almost vertical, the facial bones being thrown downwards and under the fore part of the skull. The profile of the face of the Negro, on the other hand, is singularly inclined, the front part of the jaws projecting far forward beyond the level of the fore part of the skull. In the former case the skull is said to be ' *orthognathous* '

* In no normal human skull does the breadth of the brain case exceed its length.

8

or straight-jawed; in the latter, it is called '*prognathous*,'

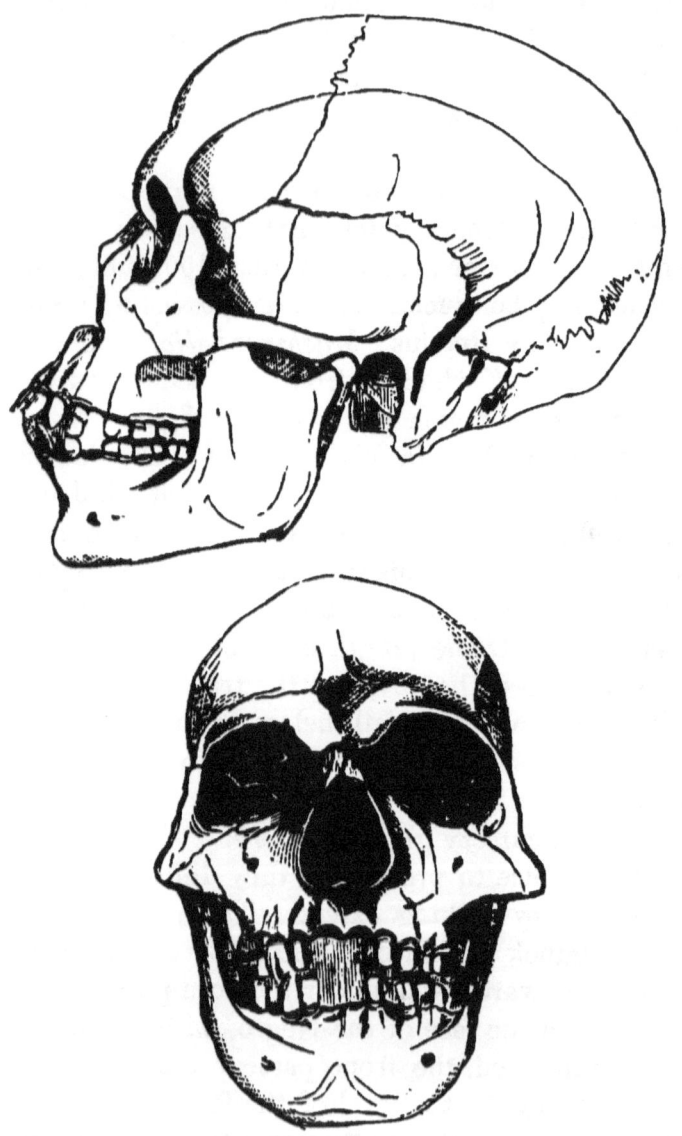

Fig. 28.—Oblong and prognathous skull of a Negro; side and front views
One-third of the natural size.

a term which has been rendered, with more force than elegance, by the Saxon equivalent,—'snouty.'

Various methods have been devised in order to express with some accuracy the degree of prognathism or orthognathism of any given skull; most of these methods being essentially modifications of that devised by Peter Camper, in order to attain what he called the 'facial angle.'

But a little consideration will show that any 'facial angle' that has been devised, can be competent to express the structural modifications involved in prognathism and orthognathism, only in a rough and general sort of way. For the lines, the intersection of which forms the facial angle, are drawn through points of the skull, the position of each of which is modified by a number of circumstances, so that the angle obtained is a complex resultant of all these circumstances, and is not the expression of any one definite organic relation of the parts of the skull.

I have arrived at the conviction that no comparison of crania is worth very much, that is not founded upon the establishment of a relatively fixed base line, to which the measurements, in all cases, must be referred. Nor do I think it is a very difficult matter to decide what that base line should be. The parts of the skull, like those of the rest of the animal framework, are developed in succession: the base of the skull is formed before its sides and roof; it is converted into cartilage earlier and more completely than the sides and roof: and the cartilaginous base ossifies, and becomes soldered into one piece long before the roof. I conceive then that the base of the skull may be demonstrated developmentally to be its relatively fixed part, the roof and sides being relatively moveable.

The same truth is exemplified by the study of the modifications which the skull undergoes in ascending from the lower animals up to man.

In such a mammal as a Beaver (Fig. 29), a line (*a. b.*) drawn through the bones, termed basioccipital, basisphe-

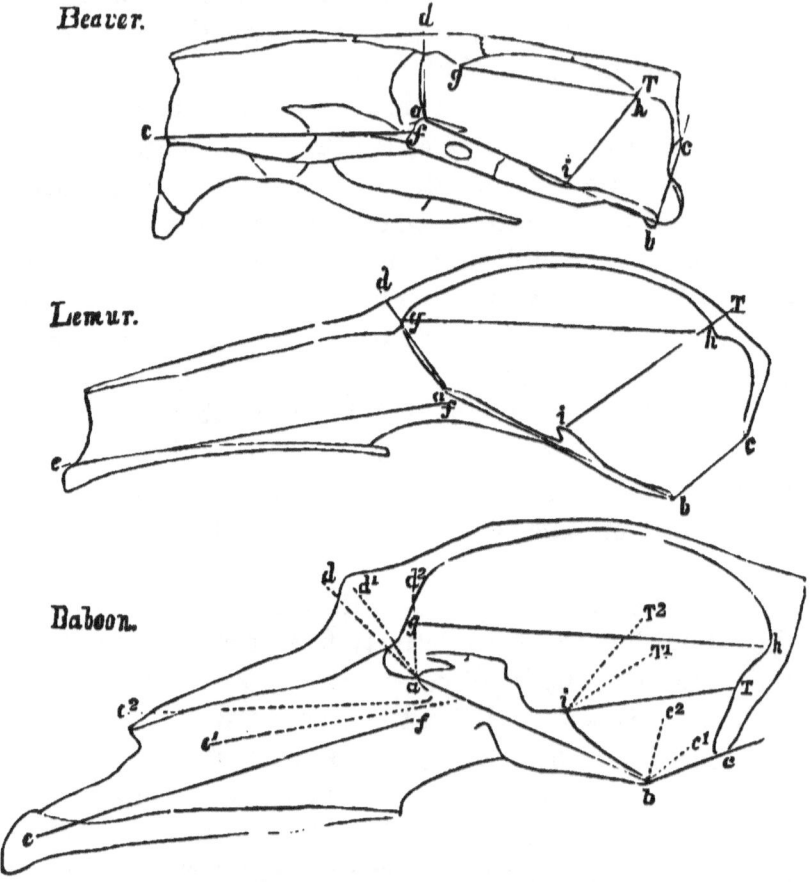

Beaver.

Lemur.

Baboon.

FIG. 29.—Longitudinal and vertical sections of the skulls of a Beaver (*Castor Canadensis*), a Lemur (*L. Catta*), and a Baboon (*Cynocephalus Papio*), *a b*, the basicranial axis; *b c*, the occipital plane; *i T*, the tentorial plane; *a d*, the olfactory plane; *f e*, the basifacial axis; *c b a*, occipital angle; *T i a*, tentorial angle; *d a b*, olfactory angle; *e f b*, cranio-facial angle; *g h*, extreme length of the cavity which lodges the cerebral hemispheres or 'cerebral length.' The length of the basicranial axis as to this length, or, in other words, the proportional length of the line *g h* to that of *a b* taken as 100, in the three skulls, is as follows:—Beaver 70 to 100; Le-

noid, and presphenoid, is very long in proportion to the
extreme length of the cavity which contains the cerebral
hemispheres (*g. h.*). The plane of the occipital foramen
(*b. c.*) forms a slightly acute angle with this 'basicranial
axis,' while the plane of the tentorium (*i. T.*) is inclined
at rather more than 90° to the 'basicranial axis'; and so
is the plane of the perforated plate (*a. d.*), by which the
filaments of the olfactory nerve leave the skull. Again,
a line drawn through the axis of the face, between the
bones called ethmoid and vomer—the "basifacial axis"
(*f. e.*) forms an exceedingly obtuse angle, where, when
produced, it cuts the 'basicranial axis.'

If the angle made by the line *b. c.* with *a. b.*, be called
the 'occipital angle,' and the angle made by the line *a. d.*
with *a. b.* be termed the 'olfactory angle,' and that made
by *i. T.* with *a. b.* the 'tentorial angle,' then all these, in
the mammal in question, are nearly right angles, varying
between 80° and 110°. The angle *e. f. b.*, or that made
by the cranial with the facial axis, and which may be
termed the 'cranio-facial angle,' is extremely obtuse,
amounting, in the case of the Beaver, to at least 150°.

But if a series of sections of mammalian skulls, intermediate beween a Rodent and a Man (Fig. 29), be examined, it will be found that in the higher crania the basicranial axis becomes shorter relatively to the cerebral
length; that the 'olfactory angle' and 'occipital angle'
become more obtuse; and that the 'cranio-facial angle,'

mur 119 to 100; Baboon 144 to 100. In an adult male Gorilla the cerebral
length is as 170 to the basicranial axis taken as 100, in the Negro (fig. 30) as
236 to 100. In the Constantinople skull (fig. 30) as 266 to 100. The cranial
difference between the highest Ape's skull and the lowest Man's is therefore
very strikingly brought out by these measurements.

In the diagram of the Baboon's skull the dotted lines *d'd²*, &c. give the
angles of the Lemur's and Beaver's skull, as laid down upon the basicranial
axis of the Baboon. The line *a b* has the same length in each diagram.

becomes more acute by the bending down, as it were, of
the facial axis upon the cranial axis. At the same time,
the roof of the cranium becomes more and more arched,
to allow of the increasing height of the cerebral hemispheres, which is eminently characteristic of man, as well
as of that backward extension, beyond the cerebellum,
which reaches its maximum in the South American monkeys. So that, at last, in the human skull (Fig. 30), the
cerebral length is between twice and thrice as great as
the length of the basicranial axis; the olfactory plane is
20° or 30° on the *under* side of that axis; the occipital
angle, instead of being less than 90°, is as much as 150°
or 160°; the cranio-facial angle may be 90° or less, and
the vertical height of the skull may have a large proportion to its length.

It will be obvious, from an inspection of the diagrams,
that the basicranial axis is, in the ascending series of
Mammalia, a relatively fixed line, on which the bones of
the sides and roof of the cranial cavity, and of the face,
may be said to revolve downwards and forwards or backwards, according to their position. The arc described by
any one bone or plane, however, is not by any means always in proportion to the arc described by another.

Now comes the important question, can we discern,
between the lowest and the highest forms of the human
cranium, anything answering, in however slight a degree,
to this revolution of the side and roof bones of the skull
upon the basicranial axis observed upon so great a scale
in the mammalian series? Numerous observations lead
me to believe that we must answer this question in the
affirmative.

The diagrams in figure 30 are reduced from very carefully made diagrams of sections of four skulls, two round
and orthognathous, two long and prognathous, taken lon-

gitudinally and vertically, through the middle. The sectional diagrams then have been superimposed, in such a manner, that the basal axes of the skulls coincide by their anterior ends and in their direction. The deviations of the rest of the contours (which represent the interior of

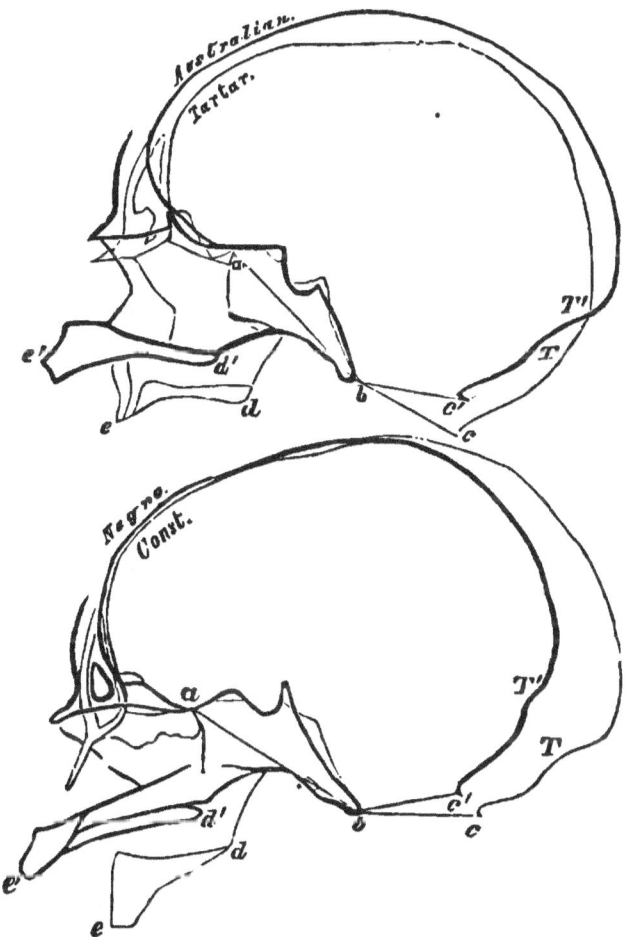

FIG. 30.—Sections of orthognathous (light contour) and prognathous (dark contour) skulls, one-third of the natural size. *a b*, Basicranial axis; *b c*, *b'c'*, plane of the occipital foramen; *d d'*, hinder end of the palatine bone; *e e'*, front end of the upper jaw; *TT'*, insertion of the tentorium.

the skulls only) show the differences of the skulls from one another when these axes are regarded as relatively fixed lines.

The dark contours are those of an Australian and of a Negro skull: the light contours are those of a Tartar skull, in the Museum of the Royal College of Surgeons; and of a well developed round skull from a cemetery in Constantinople, of uncertain race, in my own possession.

It appears, at once, from these views, that the prognathous skulls, so far as their jaws are concerned, do really differ from the orthognathous in much the same way as, though to a far less degree than, the skulls of the lower mammals differ from those of Man. Furthermore, the plane of the occipital foramen (b c) forms a somewhat smaller angle with the axis in these particular prognathous skulls than in the orthognathous; and the like may be slightly true of the perforated plate of the ethmoid—though this point is not so clear. But it is singular to remark that, in another respect, the prognathous skulls are less ape-like than the orthognathous, the cerebral cavity projecting decidedly more beyond the anterior end of the axis in the prognathous, than in the orthognathous, skulls.

It will be observed that these diagrams reveal an immense range of variation in the capacity and relative proportion to the cranial axis, of the different regions of the cavity which contains the brain, in the different skulls. Nor is the difference in the extent to which the cerebral overlaps the cerebellar cavity less singular. A round skull (Fig. 30, *Const.*) may have a greater posterior cerebral projection than a long one (Fig. 30, *Negro*).

Until human crania have been largely worked out in a manner similar to that here suggested—until it shall be an opprobrium to an ethnological collection to possess a single skull which is not bisected longitudinally—until the angles and measurements here mentioned, together with a number of others of which I cannot speak in this place, are determined, and tabulated with reference to the basicranial axis as unity, for large numbers of skulls of the different races of Mankind, I do not think we shall have any very safe basis for that ethnological craniology which aspires to give the anatomical characters of the crania of the different Races of Mankind.

At present, I believe that the general outlines of what may be safely said upon that subject may be summed up in a very few words. Draw a line on a globe from the Gold Coast in Western Africa to the steppes of Tartary. At the southern and western end of that line there live the most dolichocephalic, prognathous, curly-haired, dark-skinned of men—the true Negroes. At the northern and eastern end of the same line there live the most brachy-cephalic, orthognathous, straight-haired, yellow-skinned of men—the Tartars and Calmucks. The two ends of this imaginary line are indeed, so to speak, ethnological anti-podes. A line drawn at right angles, or nearly so, to this polar line through Europe and Southern Asia to Hindos-tan, would give us a sort of equator, around which round-headed, oval-headed, and oblong-headed, prognathous and orthognathous, fair and dark races—but none possessing the excessively marked characters of Calmuck or Negro—group themselves.

It is worthy of notice that the regions of the antipodal races are antipodal in climate, the greatest contrast the world affords, perhaps, being that between the damp, hot, steaming, alluvial coast plains of the West Coast of Africa and the arid, elevated steppes and plateaux of Central Asia, bitterly cold in winter, and as far from the sea as any part of the world can be.

8*

From Central Asia eastward to the Pacific Islands and sub-continents on the one hand, and to America on the other, brachycephaly and orthognathism gradually diminish, and are replaced by dolichocephaly and prognathism, less, however, on the American Continent (throughout the whole length of which a rounded type of skull prevails largely, but not exclusively)* than in the Pacific region, where, at length, on the Australian Continent and in the adjacent islands, the oblong skull, the projecting jaws, and the dark skin reappear ; with so much departure, in other respects, from the Negro type, that ethnologists assign to these people the special title of ' Negritoes.'

The Australian skull is remarkable for its narrowness and for the thickness of its walls, especially in the region of the supraciliary ridge, which is frequently, though not by any means invariably, solid throughout, the frontal sinuses remaining undeveloped. The nasal depression, again, is extremely sudden, so that the brows overhang and give the countenance a particularly lowering, threatening expression. The occipital region of the skull, also, not unfrequently becomes less prominent ; so that it not only fails to project beyond a line drawn perpendicular to the hinder extremity of the glabello-occipital line, but even, in some cases, begins to shelve away from it, forwards, almost immediately. In consequence of this circumstance the parts of the occipital bone which lie above and below the tuberosity make a much more acute angle with one another than is usual, whereby the hinder part of the base of the skull appears obliquely truncated. Many Australian skulls have a considerable height, quite equal to that of the average of any other race, but there

* See Dr. D. Wilson's valuable paper "On the supposed prevalence of one Cranial Type throughout the American aborigines."—Canadian Journal, Vol. II. 1857.

are others in which the cranial roof becomes remarkably depressed, the skull, at the same time, elongating so much that, probably, its capacity is not diminished. The majority of skulls possessing these characters which I have seen, are from the neighbourhood of Port Adelaide in South Australia, and have been used by the natives as water vessels; to which end the face has been knocked away, and a string passed through the vacuity and the occipital foramen, so that the skull was suspended by the greater part of its basis.

Figure 31 represents the contour of a skull of this kind from Western Port, with the jaw attached, and of the Neanderthal skull, both reduced to one third of the size of nature. A small additional amount of flattening and lengthening, with a corresponding increase of the supra-

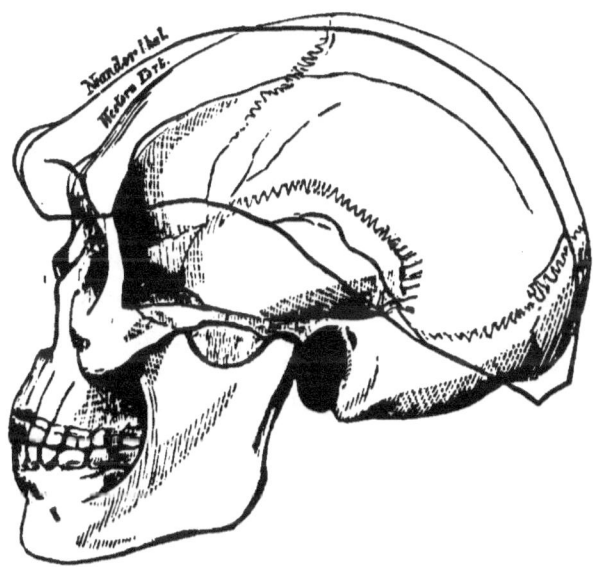

Fig. 31.—An Australian skull from Western Port, in the Museum of the Royal College of Surgeons, with the contour of the Neanderthal skull. Both reduced to one-third the natural size.

ciliary ridge would convert the Australian brain case into a form identical with that of the aberrant fossil.

And now, to return to the fossil skulls, and to the rank which they occupy among, or beyond, these existing varieties of cranial conformation. In the first place, I must remark, that, as Professor Schmerling well observed (*suprà, p.* 142) in commenting upon the Engis skull, the formation of a safe judgment upon the question is greatly hindered by the absence of the jaws from both the crania, so that there is no means of deciding, with certainty, whether they were more or less prognathous than the lower existing races of mankind. And yet, as we have seen, it is more in this respect than any other, that human skulls vary, towards and from, the brutal type—the brain case of an average dolichocephalic European differing far less from that of a Negro, for example, than his jaws do. In the absence of the jaws, then, any judgment on the relations of the fossil skulls to recent Races must be accepted with a certain reservation.

But taking the evidence as it stands, and turning first to the Engis skull, I confess I can find no character in the remains of that cranium which, if it were a recent skull, would give any trustworthy clue as to the Race to which it might appertain. Its contours and measurements agree very well with those of some Australian skulls which I have examined—and especially has it a tendency towards that occipital flattening, to the great extent of which, in some Australian skulls, I have alluded. But all Australian skulls do not present this flattening, and the supraciliary ridge of the Engis skull is quite unlike that of the typical Australians.

On the other hand, its measurements agree equally well with those of some European skulls. And assuredly, there is no mark of degradation about any part of its structure. It is, in fact, a fair average human skull, which might have belonged to a philosopher, or might have contained the thoughtless brains of a savage.

The case of the Neanderthal skull is very different. Under whatever aspect we view this cranium, whether we regard its vertical depression, the enormous thickness of its supraciliary ridges, its sloping occiput, or its long and straight squamosal suture, we meet with ape-like characters, stamping it as the most pithecoid of human crania yet discovered. But Professor Schaaffhausen states (*suprà, p.* 152), that the cranium, in its present condition, holds 1033.24 cubic centimetres of water, or about 63 cubic inches, and as the entire skull could hardly have held less than an additional 12 cubic inches, its capacity may be estimated at about 75 cubic inches, which is the average capacity given by Morton for Polynesian and Hottentot skulls.

So large a mass of brain as this, would alone suggest that the pithecoid tendencies, indicated by this skull, did not extend deep into the organization ; and this conclusion is borne out by the dimensions of the other bones of the skeleton given by Professor Schaaffhausen, which show that the absolute height and relative proportions of the limbs were quite those of an European of middle stature. The bones are indeed stouter, but this and the great development of the muscular ridges noted by Dr. Schaaffhausen, are characters to be expected in savages. The Patagonians, exposed without shelter or protection to a climate possibly not very dissimilar from that of Europe at the time during which the Neanderthal man lived, are remarkable for the stoutness of their limb bones.

In no sense, then, can the Neanderthal bones be regarded as the remains of a human being intermediate between Men and Apes. At most, they demonstrate the existence of a Man whose skull may be said to revert somewhat towards the pithecoid type—just as a Carrier, or a Pouter, or a Tumbler, may sometimes put on the plumage of its primitive stock, the *Columba livia.* And

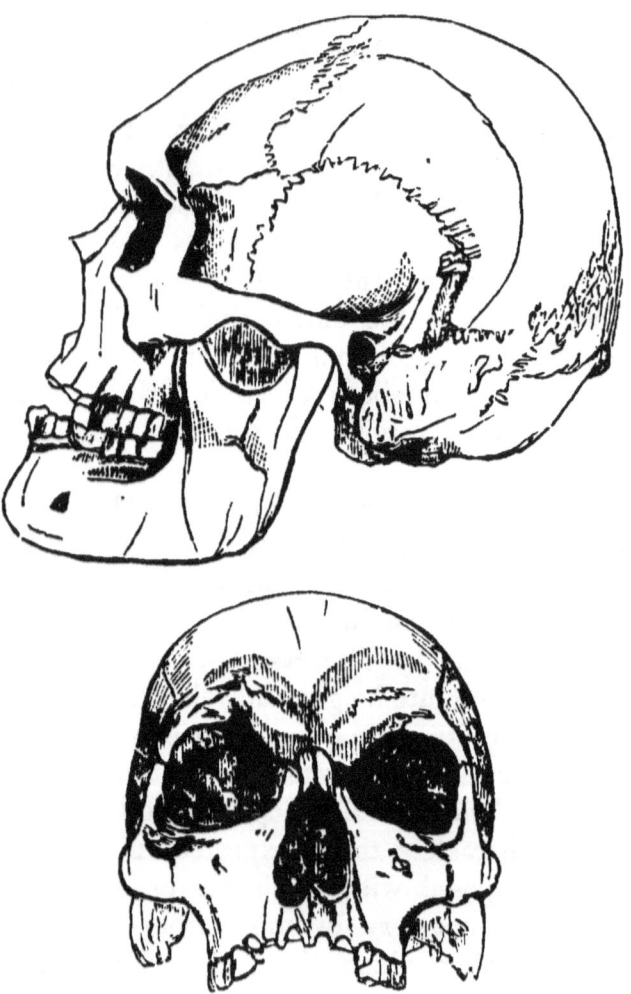

Fig. 32.—Ancient Danish skull from a tumulus at Borreby; one-third of the natural size. From a camera lucida drawing by Mr. Busk.

indeed, though truly the most pithecoid of known human skulls, the Neanderthal cranium is by no means so isolated as it appears to be at first, but forms, in reality, the extreme term of a series leading gradually from it to the highest and best developed of human crania. On the one hand, it is closely approached by the flattened Australian skulls, of which I have spoken, from which other Australian forms lead us gradually up to skulls having very much the type of the Engis cranium. And, on the other hand, it is even more closely affined to the skulls of certain ancient people who inhabited Denmark during the 'stone period,' and were probably either contemporaneous with, or later than, the makers of the 'refuse heaps,' or 'Kjokkenmöddings' of that country.

The correspondence between the longitudinal contour of the Neanderthal skull and that of some of those skulls from the tumuli at Borreby, very accurate drawings of which have been made by Mr. Busk, is very close. The occiput is quite as retreating, the supraciliary ridges are nearly as prominent, and the skull is as low. Furthermore, the Borreby skull resembles the Neanderthal form more closely than any of the Australian skulls do, by the much more rapid retrocession of the forehead. On the other hand, the Borreby skulls are all somewhat broader, in proportion to their length, than the Neanderthal skull, while some attain that proportion of breadth to length (80 : 100) which constitutes brachycephaly.

In conclusion, I may say, that the fossil remains of Man hitherto discovered do not seem to me to take us appreciably nearer to that lower pithecoid form, by the modification of which he has, probably, become what he is. And considering what is now known of the most ancient Races of men ; seeing that they fashioned flint axes and flint knives and bone-skewers, of much the same pattern as those fabricated by the lowest savages at the present day, and that we have every reason to believe the habits and modes of living of such people to have remained the same from the time of the Mammoth and the tichorhine Rhinoceros till now, I do not know that this result is other than might be expected.

Where, then, must we look for primæval Man? Was the oldest *Homo sapiens* pliocene or miocene, or yet more ancient? In still older strata do the fossilized bones of an Ape more anthropoid, or a Man more pithecoid, than any yet known await the researches of some unborn paleontologist?

Time will show. But, in the meanwhile, if any form of the doctrine of progressive development is correct, we must extend by long epochs the most liberal estimate that has yet been made of the antiquity of Man.

THE END.

www.ingramcontent.com/pod-product-compliance
Lightning Source LLC
Chambersburg PA
CBHW022356020726
47500CB00002B/299